PENGUIN BOOKS
Finding the Mother Tree

Dr Suzanne Simard was raised in the Monashee Mountains of British Columbia. She is Professor of Forest Ecology in the University of British Columbia's Faculty of Forestry, and has earned a global reputation for her research on tree connectivity and communication and its impact on the health and biodiversity of forests.

FINDING THE MOTHER TREE

Uncovering the Wisdom and Intelligence of the Forest

Suzanne Simard

PENGUIN BOOKS

PENGUIN BOOKS

UK | USA | Canada | Ireland | Australia
India | New Zealand | South Africa

Penguin Books is part of the Penguin Random House group of companies whose
addresses can be found at global.penguinrandomhouse.com.

First published in the United States of America by Alfred A. Knopf,
a division of Penguin Random House LLC 2021
First published in Great Britain by Allen Lane 2021
Published in Penguin Books 2022
003

Printed and bound in Great Britain by Clays Ltd, Elcograf S.p.A.

The authorized representative in the EEA is Penguin Random House Ireland,
Morrison Chambers, 32 Nassau Street, Dublin D02 YH68

A CIP catalogue record for this book is available from the British Library

ISBN: 978-0-141-99028-6

www.greenpenguin.co.uk

MIX
Paper from
responsible sources
FSC® C018179

Penguin Random House is committed to a
sustainable future for our business, our readers
and our planet. This book is made from Forest
Stewardship Council® certified paper.

For my daughters,

HANNAH AND NAVA

But man is a part of nature, and his war against
nature is inevitably a war against himself.

—RACHEL CARSON

CONTENTS

A FEW NOTES FROM THE AUTHOR

I use the British spelling "mycorrhizas" as the plural of "mycorrhiza" because it comes more naturally to me and may be easier for readers to recall or say. However, "mycorrhizae" is also frequently employed, especially in North America. Either plural is correct usage.

For names of species, I have used a mixture of Latin and common names throughout. For trees and plants, I usually refer to the common name at the species level, but for fungi I generally only provide the name of the genus.

I have changed the names of some people to protect their identity.

FINDING
THE
MOTHER
TREE

CONNECTIONS

For generations, my family has made its living cutting down forests. Our survival has depended on this humble trade.

It is my legacy.

I have cut down my fair share of trees as well.

But nothing lives on our planet without death and decay. From this springs new life, and from this birth will come new death. This spiral of living taught me to become a sower of seeds too, a planter of seedlings, a keeper of saplings, a part of the cycle. The forest itself is part of much larger cycles, the building of soil and migration of species and circulation of oceans. The source of clean air and pure water and good food. There is a necessary wisdom in the give-and-take of nature—its quiet agreements and search for balance.

There is an extraordinary generosity.

Working to solve the mysteries of what made the forests tick, and how they are linked to the earth and fire and water, made me a scientist. I watched the forest, and I listened. I followed where my curiosity led me, I listened to the stories of my family and people, and I learned from the scholars. Step-by-step—puzzle by puzzle—I poured everything I had into becoming a sleuth of what it takes to heal the natural world.

I was lucky to become one of the first in the new generation of women in the logging industry, but what I found was not what I had grown up to understand. Instead I discovered vast landscapes cleared of trees, soils stripped of nature's complexity, a persistent harshness of elements, communities devoid of old trees, leaving the young ones vulnerable, and an industrial order that felt hugely, terribly misguided. The industry had declared war on those parts of the ecosystem— the leafy plants and broadleaf trees, the nibblers and gleaners and infesters—that were seen as competitors and parasites on cash crops but that I was discovering were necessary for healing the earth. The whole forest—central to my being and sense of the universe—was suffering from this disruption, and because of that, all else suffered too.

I set out on scientific expeditions to figure out where we had gone so very wrong and to unlock the mysteries of why the land mended itself when left to its own devices—as I'd seen happen when my ancestors logged with a lighter touch. Along the way, it became uncanny, almost eerie, the way my work unfolded in lockstep with my personal life, entwined as intimately as the parts of the ecosystem I was studying.

The trees soon revealed startling secrets. I discovered that they are in a web of interdependence, linked by a system of underground channels, where they perceive and connect and relate with an ancient intricacy and wisdom that can no longer be denied. I conducted hundreds of experiments, with one discovery leading to the next, and through this quest I uncovered the lessons of tree-to-tree communication, of the relationships that create a forest society. The evidence was at first highly controversial, but the science is now known to be rigorous, peer-reviewed, and widely published. It is no fairy tale, no flight of fancy, no magical unicorn, and no fiction in a Hollywood movie.

These discoveries are challenging many of the management practices that threaten the survival of our forests, especially as nature struggles to adapt to a warming world.

My queries started from a place of solemn concern for the future of our forests but grew into an intense curiosity, one clue leading to another, about how the forest was more than just a collection of trees.

In this search for the truth, the trees have shown me their perceptiveness and responsiveness, connections and conversations. What started as a legacy, and then a place of childhood home, solace, and adventure in western Canada, has grown into a fuller understanding of the intelligence of the forest and, further, an exploration of how we can regain our respect for this wisdom and heal our relationship with nature.

One of the first clues came while I was tapping into the messages that the trees were relaying back and forth through a cryptic underground fungal network. When I followed the clandestine path of the conversations, I learned that this network is pervasive through the *entire* forest floor, connecting all the trees in a constellation of tree hubs and fungal links. A crude map revealed, stunningly, that the biggest, oldest timbers are the sources of fungal connections to regenerating seedlings. Not only that, they connect to all neighbors, young and old, serving as the linchpins for a jungle of threads and synapses and nodes. I'll take you through the journey that revealed the most shocking aspect of this pattern—that it has similarities with our own human brains. In it, the old and young are perceiving, communicating, and responding to one another by emitting chemical signals. *Chemicals identical to our own neurotransmitters. Signals created by ions cascading across fungal membranes.*

The older trees are able to discern which seedlings are their own kin.

The old trees nurture the young ones and provide them food and water just as we do with our own children. It is enough to make one pause, take a deep breath, and contemplate the social nature of the forest and how this is critical for evolution. The fungal network appears to wire the trees for fitness. And more. These old trees are mothering their children.

The Mother Trees.

When Mother Trees—the majestic hubs at the center of forest communication, protection, and sentience—die, they pass their wisdom to their kin, generation after generation, sharing the knowledge of what helps and what harms, who is friend or foe, and how to adapt and survive in an ever-changing landscape. It's what all parents do.

How is it possible for them to send warning signals, recognition messages, and safety dispatches as rapidly as telephone calls? How do they help one another through distress and sickness? Why do they have human-like behaviors, and why do they work like civil societies?

After a lifetime as a forest detective, my perception of the woods has been turned upside down. With each new revelation, I am more deeply embedded in the forest. The scientific evidence is impossible to ignore: the forest is wired for wisdom, sentience, and healing.

This is not a book about how we can save the trees.

This is a book about how the trees might save us.

GHOSTS IN THE FOREST

I was alone in grizzly country, freezing in the June snow. Twenty years old and green, I was working a seasonal job for a logging company in the rugged Lillooet Mountain Range of western Canada.

The forest was shadowed and deathly quiet. And from where I stood, full of ghosts. One was floating straight toward me. I opened my mouth to scream, but no sound emerged. My heart lodged in my throat as I tried to summon my rationality—and then I laughed.

The ghost was just heavy fog rolling through, its tendrils encircling the tree trunks. No apparitions, only the solid timbers of my industry. The trees were *just trees*. And yet Canadian forests always felt haunted to me, especially by my ancestors, the ones who'd defended the land or conquered it, who came to cut, burn, and farm the trees.

It seems the forest always remembers.

Even when we'd like it to forget our transgressions.

It was midafternoon already. Mist crept through the clusters of sub-alpine firs, coating them with a sheen. Light-refracting droplets held entire worlds. Branches burst with emerald new growth over a fleece of jade needles. Such a marvel, the tenacity of the buds to surge with life every spring, to greet the lengthening days and warming weather with exuberance, no matter what hardships were brought by winter. Buds encoded to unfold with primordial leaves in tune with the fair-ness of previous summers. I touched some feathery needles, comforted by their softness. Their stomata—the tiny holes that draw in carbon dioxide to join with water to make sugar and pure oxygen—pumped fresh air for me to gulp.

Nestled against the towering, hardworking elders were teenaged saplings, and leaning into them were even younger seedlings, all huddling as families do in the cold. The spires of the wrinkled old firs stretched skyward, sheltering the rest. The way my mother and father, grandmothers and grandfathers protected me. Goodness knows, I'd needed as much care as a seedling, given that I was always getting into trouble. When I was twelve, I'd crawled along a sweeper tree leaning over the Shuswap River to see how far out I could go. I tried to retreat but slipped and fell into the current. Grampa Henry jumped into his hand-built riverboat and grabbed my shirt collar right before I would have disappeared into the rapids.

Snow lay deeper than a grave nine months of the year here in the mountains. The trees far outmatched me, their DNA forged so they'd thrive despite the extremes of an inland climate that would chew me up and spit me out. I tapped a limb of an elder to show gratitude for its hovering over vulnerable offspring and nestled a fallen cone in the crook of a branch.

I pulled my hat over my ears while stepping off the logging road and waded deeper into the forest through the snow. Despite it being only a few hours before darkness, I paused at a log, a casualty of saws that had cleared the road right-of-way. The pale round face of its cut end showed age rings as fine as eyelashes. The blond-colored earlywood, the spring cells plump with water, were edged by dark-brown cells of latewood formed in August when the sun is high and drought settles in. I counted the rings, marking each decade with a pencil—the tree was a couple hundred years old. Over twice the number of years my own family had lived in these forests. How had the trees weathered the changing cycles of growth and dormancy, and how did this compare to the joys and hardships my family had endured in a fraction of the time? Some rings were wider, having grown plenty in rainy years, or perhaps in sunny years after a neighboring tree blew over, and others were almost too narrow to see, having grown slowly during a drought, a cold summer, or some other stress. These trees persisted through climatic upheavals, suffocating competition, and ravaging fire, insect, or wind disruptions, far eclipsing the colonialism, world wars, and the

Camping at Shuswap Lake near Sicamous, British
Columbia, 1966. Left to right: Kelly, three; Robyn, seven;
and Mum, Ellen June, twenty-nine; I'm five. We arrived
in our 1962 Ford Meteor after barely escaping a rockslide
on the Trans-Canada Highway; rocks flew down the
mountain straight through the car window and landed
on Mum's lap.

dozen or so prime ministers my family had lived through. They were
ancestors to my ancestors.

A chattering squirrel ran along the log, warning me away from his
cache of seeds at the base of the stump. I was the first woman to work
for the logging company, an outfit that was part of a rough, dangerous
business starting to open its doors to the occasional female student.
The first day on the job, a few weeks back, I'd visited a clear-cut—a
complete felling of trees in a thirty-hectare patch—with my boss, Ted,
to check that some new seedlings had been planted according to government rules. He knew how a tree should and should not be planted,
and his low-key approach kept workers going through their exhaustion. Ted had been patient with my embarrassment at not knowing a

Temperate rain forest typical of Mum's and Dad's childhood homes in
British Columbia

J-root from a deep plug, but I'd watched and listened. Soon enough,
I was entrusted with the job of assessing established plantations—
seedlings put in to replace harvested trees. I wasn't about to screw up.

Today's plantation awaited me beyond this old forest. The company had chopped down a large parcel of velvety old subalpine firs
and planted prickly needled spruce seedlings this last spring. My task
was to check the progress of those new growths. I hadn't been able to

take the logging road into the clear-cut because it had been washed out—a gift, since I could detour past these mist-wrapped beauties, but I stopped at a massive pile of fresh grizzly scat.

Fog still draped the trees, and I could have sworn something was sliding along in the distance. I looked harder. It was the pale green trusses of the lichen called old man's beard because of the way it sways from branches. Old lichen that particularly thrived on old trees. I plunged the button on my air horn to warn off the specter of bears. I'd inherited my fear of them from my mother, who was a child when her grandfather, my great-grampa Charles Ferguson, shot and killed one that was inches from mauling her on the porch. Great-Grampa Charles was a turn-of-the-twentieth-century pioneer in Edgewood, an outpost in the Inonoaklin Valley along the Arrow Lakes of the Columbia basin in British Columbia. With axes and horses, he and his wife, Ellen, cleared the Sinixt Nation land they had homesteaded to grow hay and tend cattle. Charles was known to wrestle with bears and shoot wolves that tried to kill his chickens. He and Ellen raised three children: Ivis, Gerald, and my grandmother Winnie.

I crawled over logs covered with moss and mushrooms, inhaling the evergreen mist. One had a river of tiny *Mycena* mushrooms flowing along the cracks down its length before fanning along a splay of tree roots that dwindled to rotten spindles. I'd been puzzling over what roots and fungi had to do with the health of forests—the harmony of things large and small, including concealed and overlooked elements. My fascination with tree roots had started from my growing up amazed at the irrepressible power of the cottonwoods and willows my parents had planted in our backyard when their massive roots cracked the foundation of our basement, tilted over the doghouse, and heaved up our sidewalk. Mum and Dad fell into worried discussions of what to do with the problem they'd unwittingly created in our little plot of land in trying to reconstruct the feel of trees surrounding their own childhood homes. I'd watched in awe each spring as a multitude of germinants emerged from cottony seeds amid halos of mushrooms fanning around the base of the trees, and I'd become horrified, at eleven, when the city ran a pipeline spewing foamy water into the river beside my house, where the effluent killed the cottonwoods along the shore. First the tops of the crowns thinned, then black cankers appeared around the furrowed trunks, and by the next spring the great trees were dead. No new germinants got established among the yellow outflow. I wrote to the mayor, and my letter went unanswered.

I picked one of the tiny mushrooms. The bell-shaped elf caps of the *Mycena* were dark brown at the apex and faded into translucent yellow at the margins, revealing gills underneath and a fragile stem. The stipes—stems—were rooted in the furrows of the bark, helping the log decay. These mushrooms were so delicate it seemed impossible they could decompose a whole log. But I knew they could. Those dead cottonwoods along the riverbank in my childhood had fallen and sprouted mushrooms along their thin, cracking skin. Within a few years, the spongy fibers of decayed wood had completely disappeared into the ground. These fungi had evolved a way to break down wood by exuding acids and enzymes and using their cells to absorb the wood's energy and nutrients. I launched off the log, landed with my caulk spikes in the duff, and grabbed clumps of fir saplings to leverage myself up the slope. The saplings had found a spot to capture a balance between the light of the sun and the wetness of the snowmelt.

A *Suillus* mushroom—tucked near a seedling that had established a few years back—was wearing a scaly brown pancake cap over a yellow porous underbelly and a fleshy stem that disappeared into the ground.

Pancake mushrooms (*Suillus lakei*)

In a burst of rain, the mushroom had sprung out of the dense network of branching fungal threads running deep through the forest floor. Like a strawberry fruiting from its vast, intricate system of roots and runners. With a boost of energy from the earthen threads, the fungal cap had unfurled like an umbrella, leaving traces of a lacy veil hugging the brown-spotted stem about halfway up. I picked the mushroom, this fruit of the fungus that otherwise lived mainly belowground. The cap's underside was like a sundial of radiating pores. Each oval-shaped opening housed minuscule stalks built to discharge spores like sparks from a firecracker. Spores are the "seeds" of fungi, full of DNA that binds, recombines, and mutates to produce novel genetic material that is diverse and adapted for changing environmental conditions. Sprinkled around the colorful cavity left by the picking was a halo of cinnamon-brown spores. Other spores would have caught an updraft, latched on to the legs of a flying insect, or become the dinner of a squirrel.

Extending downward in the tiny crater still holding the remains of the mushroom's stem were fine yellow threads, the strands braiding into an intricately branching veil of fungal *mycelium,* the network that blankets the billions of organic and mineral particles making up the soil. The stem bore broken threads that had been part of this web before I ungraciously ripped it from its moorings. The mushroom is the visible tip of something deep and elaborate, like a thick lace tablecloth knitted into the forest floor. The threads left behind were fanning through the litter—fallen needles, buds, twigs—searching for, entwining with, and absorbing mineral riches. I wondered whether this *Suillus* mushroom might be a type of decay fungus like the *Mycenas,* a rotter of wood and litter, or if it had some other role. I stuck it into my pocket along with the *Mycena.*

The clear-cut where the seedlings replaced the chopped-down trees was still not visible. Dark clouds were gathering, and I pulled my yellow rain jacket out of my vest. It was worn from bushwhacking and not as waterproof as it should have been. Each step farther from the truck added to an aura of danger and my foreboding that I wouldn't be on the road by nightfall. But I'd inherited an instinct for pushing through hardship from Grannie Winnie, a teenager when her mother, Ellen, succumbed to the flu in the early 1930s. The family was snowed in and bedridden, with Ellen dead in her room, when the neighbors finally broke through the frozen valley and chest-deep snow to check on the Ferguson clan.

My boot slipped, and I grabbed a sapling, which came loose in my hand as I tumbled down the pitch, flattening other saplings before coming to rest against a sodden log, still clutching the octopus of jagged roots. The young tree looked to be a teenager, the whorls of lateral branches demarcating each year adding up to about fifteen. A rain cloud started to spit, soaking my jeans. Drops beaded on the oilskin of my scruffy jacket.

There was no room for weakness on this job, and I'd cultivated a tough exterior in a boy's world for as long as I could remember. I wanted to be as good as my younger brother, Kelly, and the ones who had Québécois names like Leblanc and Gagnon and Tremblay, so I learned to play street ice hockey with the neighborhood gang when the temperature was minus twenty. I played goalie, the least coveted position. They took hard shots at my knees, but I kept my black-and-blue legs concealed under my jeans. The way Grannie Winnie kept on as best she could, resuming her job of galloping her horse through the Inonoaklin Valley, delivering mail and flour to the homesteads, soon after her mother died.

I stared at the clump of roots in my fist. Clinging to them was glistening humus that reminded me of chicken manure. Humus is the greasy black rot in the forest floor sandwiched between the fresh litter from fallen needles and dying plants above and the mineral soil weathered from bedrock below. Humus is the product of plant decay. It's where the dead plants and bugs and voles are buried. Nature's compost. Trees love to root in the humus, not so much above or below it, because there they can access the bounty of nutrients.

But these root tips were glowing yellow, like lights on a Christmas tree, and they ended in a gossamer of mycelium of the same color. The threads of this streaming mycelium looked close to the same color as those radiating into the soil from the stems of the *Suillus* mushrooms, and from my pocket I took out the one I'd picked. I held the clump of root tips with its cascading yellow gossamer in one hand and the *Suillus* mushroom with its broken mycelium in the other. I studied them closely, but I could not tell them apart.

Winnifred Beatrice Ferguson (Grannie Winnie) at the Ferguson farm in Edgewood, British Columbia, ca. 1934, when she was twenty years old, shortly after her mum died. Winn carried on raising the chickens, milking the cows, and pitching the hay. She rode her horse like the wind and shot a bear out of the apple tree. Grannie rarely spoke of her mum, but on my last walk with her along the waterfront of Nakusp, when she was eighty-six years old, she cried to me, "I miss my mum."

Maybe *Suillus* was a friend of the roots, not a decomposer of dead things as *Mycena* was? My instinct has always been to listen to what living things are saying. We think that most important clues are large, but the world loves to remind us that they can be beautifully small. I began to dig into the forest floor. The yellow mycelium seemed to coat every minuscule particle of soil. Hundreds of miles of threads running under my palms. No matter the lifestyle, these fungal branching filaments, called *hyphae*—along with the mushroom fruit they spawned—appeared to be only a smattering of the vast mycelium in the soil.

My water bottle was in the back zipped pocket of my vest, and I washed the soil crumbs from the rest of the root tips. I'd never seen such a rich bouquet of fungus—certainly not this brilliant a yellow, plus white and pink too—each color wrapped around a separate tip, bearded with gossamer. Roots need to reach far and in awkward spaces for nutrients. But why were so many fungal threads not only sprouting from the root tips but blazing with a palette like this? Was each color a different fungal species? Did each do a different job in the soil?

I was in love with this work. The rush of excitement climbing through this majestic glade was far more intense than my fear of bears or ghosts. I set the roots of my ripped-out seedling, with their vivid netting of fungus, near a guardian tree. The seedlings had shown me the textures and tones of the forest's underworld. Yellows and whites and shades of dusty pink that reminded me of the wild roses I grew up with. The soil where they had found purchase was like a book, one colorful page layered on the next, each unfolding the story of how everything was nourished.

When I finally made it into the clear-cut, I squinted in the glare filtering through the drizzle. I knew what to expect, but my heart still jolted. Every tree had been cut down to a stump. White bones of wood jutted out of the soil. Weathered by the wind and rain, the last scraps of bark sloughed onto the ground. I picked my way past severed limbs, feeling the pain of their neglect. I lifted a branch to uncover a young tree, just as I'd picked garbage off the flowers trying to bloom under the trash piles in the hills above the neighborhood when I was a child. I knew the importance of these gestures. Some little velvety firs had been orphaned near the stumps of their parents and were trying to recover from the shock of their loss. Their recuperation would be arduous given the slow shoot growth since the harvest. I touched the tiny terminal bud of the one closest to me.

Some white-flowered rhododendrons and huckleberry shrubs had also ducked the zip of the saw. I was a part of this harvesting of lumber, this business of chopping down trees to clear the spaces where they were free, wild, whole. My colleagues were drawing up plans for the next clear-cuts, to keep the mill going and their families fed, and I understood this need too. But the saws wouldn't stop until whole valleys were gone.

I walked toward seedlings in a crooked line amid the rhododendrons and huckleberries. The crew that had done the planting to replace the harvested elder firs had inserted prickly spruce seedlings, now ankle high. It might seem odd not to replace the subalpine firs they'd taken down with more subalpine firs. But spruce wood is more valuable. It's tightly grained, resistant to decay, and coveted for high-grade lumber. Mature subalpine-fir timber is weak and punky.

The government also encouraged planting the seedlings in garden-like rows to ensure no patch of soil was left bare. This was because timber grown in grids of evenly spaced trees yielded more wood than scattered clumps. At least in theory. By filling in all the gaps, they figured they could grow more wood than occurred naturally. With every corner chockablock, they felt justified in bigger harvests, in anticipation of future yields. And logical rows made everything more countable. Same rationale as my Grannie Winnie planting her garden in rows, but she worked the soil and varied her crops over the years.

The first spruce seedling I checked was alive, but barely, with yellowish needles. Its spindly stem was pathetic. How was it supposed to survive this brutal terrain? I looked up the planted row. All the new seedlings were struggling—every single sad little planting. Why did they look so *awful*? Why, in contrast, did the wild firs germinating in that old-growth patch look so *brilliant*? I pulled out my field book, wiped needles off the waterproof cover, and cleaned my glasses. The replanting was supposed to heal what we'd taken, and we were failing miserably. What prescription should I write? I wanted to tell the company to start over again, but that expense would be frowned upon. I caved to my fears of a rebuttal and jotted, "Satisfactory, but replace the seedlings that have died."

I picked up a piece of bark shading a seedling and flicked it into the shrubs. Using a makeshift envelope fashioned from drafting paper, I collected the seedling's yellow needles. I was grateful to have my own desk in an alcove set off from the map tables and boisterous offices where men made deals and negotiated timber prices and logging costs; decided what patches of forest to cut next; awarded contracts like banner ribbons at a track meet. In my tiny space, I could work on the plantation problems in a secluded peace. Maybe the seedling's symptoms would be easy to find in the reference books, since yellowing can be caused by myriad problems.

I tried to find any seedlings that were healthy, but to no avail. What was triggering the sickness? Without a correct diagnosis, the replacement seedlings would likely suffer too.

I kicked myself for glossing over the problem, taking the easy way out for the company. The plantation was a mess. Ted would want to know if we were failing to meet the government requirements for reforestation at this site, because not succeeding meant a financial loss. He was focused on meeting the basic regeneration regulations at minimal cost, but I didn't even know what to suggest. I pulled another spruce seedling from its planting hole, wondering if the answer might be in the roots, not the needles. They had been buried tightly in the granular soil, where it was still moist in late summer. Perfect planting job. The forest floor scraped away, the planting hole plunged into the damp mineral earth below. Just as instructed. By the book. I inserted the roots back into the hole and checked another seedling. And another. Every one of them packed exactly right in a slit made by a shovel and backfilled to eliminate the air gaps, but the root plugs looked embalmed, as if they'd been shoved into a tomb. Not a single root seemed to get what it was supposed to do. None was sprouting new white tips to forage in the ground. The roots were coarse, black, and plunging straight to nowhere. The seedlings shed yellow needles because they were starving for *something*. There was an utter, maddening disconnect between the roots and the soil.

By chance a healthy subalpine fir had regenerated from a seed nearby, and I uprooted it to compare. Unlike the planted spruce, which I'd plucked like a carrot out of the soil, these sprawling fir roots were anchored so tightly that I had to plant both feet on either side of the stem and pull with all my might. The roots finally ripped out of the earth and—a parting shot—sent me stumbling. The deepest root tips had refused to unglue from the soil, no doubt in protest. But I brushed the humus and loose dirt off the torn roots I'd claimed, pulled out my water bottle, and rinsed off the remaining crumbs. Some of the root ends were like the fine tips of needles.

I was amazed to see the same bright yellow fungal threads wrapped around the root tips as I'd seen in the old-growth forest, once again exactly the same color as the mycelium, the network of fungal hyphae growing out of the stems of the *Suillus* pancake mushrooms. Digging a little more around my fir excavation, I found the yellow threads infusing the organic mat that capped the soil, forming a network of mycelium that was radiating farther and farther afield.

But what exactly were these branching fungal threads, and what were they doing? They might be beneficial hyphae meandering through the soil to pick up nutrients to deliver to the seedlings in exchange for energy. Or they could be pathogens infecting and feeding off the roots, causing vulnerable seedlings to turn yellow and die. The *Suillus* mushrooms might be popping out of the subterranean fabric to spread spores when times were good.

Or maybe these yellow threads weren't connected to *Suillus* mushrooms at all and were instead from a different fungal species. More than a million exist on earth, about six times the number of plant species, with only about 10 percent of fungal species identified. With my scant knowledge, my chances of identifying the species of these yellow threads felt like a long shot. If the threads or the mushrooms didn't hold clues, there could be other reasons why the new spruce plantings didn't flourish here.

I erased my "satisfactory" note and jotted that the plantation was a failure. A complete replanting using the same kinds of seedlings and methods—shovel planting one-year-old plug stock that is mass-produced in nurseries—felt like the cheapest way for the company to go, but not if we had to keep returning because of the same dismal result. Something different needed to be done to re-establish this forest, but what?

Plant subalpine fir? No nurseries had it available for planting, and it wasn't considered a future cash crop. We could plant spruce seedlings with bigger root systems. But the roots would still die if they couldn't sprout strong new tips. Or we could plant them so their roots touched the yellow fungal web in the soil. Maybe the yellow gossamer would keep my seedlings healthy. But the rules required that the roots be planted in the underlying granular mineral soil, not the humus—figuring that the grains of sand, silt, and clay held more water late in the summer and therefore offered a better chance of survival—and the fungus mainly lived in the humus. Water, it was thought, was the most crucial resource that soils needed to supply roots so seedlings would survive. There seemed a very low chance of a change in policy so we could plant the roots in a way that they could reach the yellow fungal threads.

I wished I had someone to talk to out here in the forest, to debate my growing sense that the fungus might be a trustworthy helper to the seedlings. Did the yellow fungus contain some secret ingredient that I—and everyone—had somehow missed?

If I didn't find an answer, I'd be haunted by turning this clear-cut into a killing field, a graveyard of tree bones. A brush field of rho-dodendrons and huckleberries instead of a new forest, a burgeoning problem, one plantation dying after another. I couldn't let this happen. I had seen forests grow back naturally after my family had logged near my home and knew it was possible for a forest to recover from a harvest. Perhaps it was because my grandparents had cut only a few trees in a stand, opening gaps where nearby cedars and hemlocks and firs could readily seed in, the new plants easily connecting to the soil. I squinted to spot the timber edge, but it was too distant. These clear-cuts were huge, and perhaps their size was part of the problem. If they had healthy roots, surely trees could regenerate in this expanse. So far, though, my job consisted of overseeing plantations with little chance of turning into anything resembling the towering cathedrals once here.

That's when I heard the grunt. Steps away, feeding on a shifting bank of blue, purple, and black berries, was a mother bear. The silver-tipped fur on the nape of her neck declared *grizzly*. A tawny cub, as tiny as Winnie-the-Pooh but with outsized fuzzy ears, was stuck to her as if she were a glue pot. The cub looked at me with soft black eyes and a glistening nose as if he wanted to run into my arms, and I smiled. But only for a moment. Mama Bear roared, and we locked eyes, both of us surprised. She towered onto her hind legs as I stood stock-still.

I was alone in the back forty with a startled grizzly. When I blew my air horn—*aaaanw!*—she only stared harder. Was I supposed to stand tall or curl into a ball? One response was to deal with black bears, and the other was for grizzlies. Why hadn't I listened to those instructions carefully?

The mama sank onto all fours, shaking her head, her chin grazing the huckleberry bushes. She nudged her little one, and they both turned on their heels. I slowly backed up as they crashed through the shrubs. She sent her cub up a tree, scrabbling on the bark. Her instinct was to protect her child.

I raced downhill toward the old forest, leaping over seedlings and rivulets, dodging the skeleton stumps of the beheaded trees, trampling shoots of hellebore and fireweed. The plants blurred into a green wall. I couldn't hear anything but my lungs grasping oxygen as I hurdled over the decaying logs, one after the other, before I spotted the company truck next to a tree just off the road, as if it had rolled to a crooked stop.

The vinyl seats were torn, and the stick shift was wobbly. I fired the ignition, threw the clutch into gear, and hit the gas. The wheels spun, but the truck didn't move. Throwing the gear into reverse made them dig deeper. I was wedged in a mudhole.

I got on the radio. "Suzanne calling Woodlands, over."

Nothing.

As darkness fell, I sent a last plea over the airwaves. A bear could easily break a window with one swing of its paw. For hours I tried to stay a waking witness to my demise, but I dozed on and off, and in between I thought about my mother's skill with escapes. I pretended she was tucking me under blankets as she used to do before we drove over the Monashee Mountains to my grandparents' house, setting a pot on my lap and brushing my blond bangs aside because I had a habit of getting carsick. "Robyn, Suzie, Kelly, get some sleep," she would whisper, set to wind in and out of ravines slicing the mountain pass. "We'll be at Grannie Winnie and Grampa Bert's soon." Summers meant a break from teaching school and her marriage. My brother and sister and I loved those days, roaming the woods away from the silent feuds of our parents. Disputes over money, about who was responsible for what, about us. Kelly in particular was happier on those escapes, tagging behind Grampa Bert picking huckleberries, or fishing with him from the government wharf, or driving to the dump where the bears scavenged. He'd listen wide-eyed to Grampa's stories of courting Grannie when he came to buy cream from the Ferguson ranch, of helping Charlie Ferguson with calving in early spring, and of filling gut wagons with cow and pig offal during the fall slaughter.

I woke with a start in the dark, neck sore, not sure where I was, the windshield opaque with my condensed breath. Wiping drizzle off the glass with the cuff of my jacket, I peered into the black for wild eyes and glanced at my watch—four a.m. Grizzlies are most active at dusk and dawn, so I checked the door locks again. Leaves rustled like a wraith creeping by. I dozed until a fierce banging on glass made me scream. A man was shouting through the foggy windshield, and I was relieved the timber company had sent Al. His border collie, Rascal,

Left to right: me, five; Mum, twenty-nine; Kelly, three; Robyn, seven; and Dad, thirty, at Grannie Winnie and Grampa Bert's house in Nakusp, ca. 1965. All of our holidays were spent either with my maternal grandparents in Nakusp or my paternal grandparents at Mabel Lake.

jumped up and scratched my door, barking. I rolled down the window to prove I was still whole.

"You okay?" Al's voice was as loud as he was marvelously tall. He was still trying to figure out how to talk with a girl forester, to do his best to include me as one of the guys. "Must have been black as molasses out here."

"It was okay," I lied.

We'd more or less succeeded in pretending it was just another night on the job, and I cracked open the door so Rascal could squeeze in for me to pet him. I loved it when Al and Rascal drove me home from work, and Al would lean out and bark at the chasing dogs, which always yelped and ran the other way, much to his delight. I found this extremely funny, which egged him on to bark even louder.

I stretched my limbs outside the truck, and Al handed me a thermos of coffee while he took a stab at driving out of the mudhole. He turned the starter, and the cold-as-a-frog engine groaned. Dew speckled the rusty hood and the pink-blossomed fireweed lined the road. Watching through the coffee's steam, I wondered if we would have to abandon the *tacot rouillé*. But the truck started on the third try. Al floored the gas pedal, and the wheels spun in place.

"Did you lock the hubs?" he asked. The hubs were dials in the middle of the front wheels, at each end of the front axle. Manually twisting them ninety degrees locked the wheels to the axle so that they, along with the back wheels, would be torqued by the engine. With all four wheels turning, the truck could plough through anything. But with the front hubs unlocked, the truck had as much traction as a cat on linoleum. I almost died when he jumped out, twisted the hubs, and drove clear of the bog. Grinning, Al handed me the keys.

"Oops," I said, banging the heel of my hand against my head.

"Don't worry, Suzanne, it happens," he said, looking down to spare me the humiliation. "It's happened to me."

I nodded. A rush of gratitude flooded through me as I followed him out of the valley.

BACK AT THE MILL, I walked rumpled and sheepish into the office, expecting to be teased, telling myself I could take it. The men glanced up, then did me the courtesy of returning immediately to chatting, enjoying the hell out of their stories of building roads, installing culverts, planning cut-blocks, cruising timber. I wondered what they thought of me, so different from the women of the town and the girls on the pinup calendar by the drafting tables, but they mostly went about their business and let me be.

I caught up with Ted a short while later, leaning against his door-jamb until he looked up. His desk was stacked with planting prescriptions and seedling orders. He had four daughters, all under the age of ten. He leaned back in his swivel chair and said with a grin, "Well, look what the cat dragged in." I knew this meant he was glad I was back safely. They'd been worried. Plus—even more crucial—our sign advertised "216 accident-free days," and I'd never hear the end of it if I'd broken our streak. When he suggested I go home, I said I had a little work to do.

I spent the day writing up my planting reports before mailing my envelope of yellow needles to the government lab to have the nutrition levels analyzed and checking the office for reference volumes about mushrooms. There were plenty of resources about logging, but books on biology were scarce as hens' teeth. I called the town library, glad to learn there was a mushroom reference guide on their shelves. At five o'clock, Ted and the guys prepared to head out to watch the football game at the Reynolds Pub before going home to their families.

"Want to join us?" he asked. Hanging out with guffawing men was the last thing I wanted, but I appreciated the gesture. He looked relieved when I thanked him and said I needed to get to the library before it closed.

I collected the mushroom book and filed my report on the plantation but vowed to keep my observations quiet and do my homework. I often feared I'd been hired into the men's club as a token of changing times, and my goose would be cooked if I came up with a half-baked idea about how mushrooms or pink or yellow quilts of fungus on roots affected seedling growth.

Kevin, another summer student, hired to help the engineers lay roads into unspoiled valleys, appeared at my desk as I gathered up my cruiser vest. He and I had become friends at the university, and we were grateful for these bush jobs. "Let's go to the Mugs'n'Jugs," he offered. It was at the other end of town from the Reynolds, and we could avoid the older guys.

"I'd love to." Hanging out with other forestry students was easy. I lived with four of them in the company bunkhouse, where I had my own dingy room with a single mattress on the floor. None of us was good at cooking, so pub nights were common. The bar also was a welcome respite because I was still hurting from breaking up with my first real love. He had wanted me to quit school and have children, but I wanted to become *someone,* my eye on a bigger prize.

At the pub, Kevin ordered a pitcher and burgers while I hunted on the jukebox for the Eagles song about taking things easy and watched the arm pick up the forty-five. When the beer came, he poured me a glass.

"They're sending me up to Gold Bridge to lay out road next week," he said. "I'm worried they'll use the beetle infestation as an excuse to cut the lodgepole-pine forests."

"Yeah, I wouldn't doubt it." I looked around to make sure no one was listening. Other students were laughing at a nearby table, downing beers, getting up to throw darts. The pub's interior was like a log cabin and smelled of mildly rotting pine. This was a company town. I blurted, "I felt like I could have friggin' died out there last night."

"Hey, you were lucky it wasn't colder. It was good the truck stalled, because you'd have been in worse trouble driving in the dark over those roads. We were trying to warn you to stay put, but I guess your radio was busted," said Kevin, back-arming foam off his mustache— someone must hand those out the moment a man opts for a life in the woods.

"I was pretty spooked," I confessed. "But at least I got to see a sweet side to Al."

"We all felt bad for you. But we knew you'd figure out how to be safe."

I smiled. He was comforting me, making me feel valued, part of the team. "New Kid in Town" sailed from the jukebox, a little mournful. In the end, I'd been protected by the powerful grip of forest mud, saving me from ghosts, bears, my nightmares.

I was born to the wild. I come from the wild.

I can't tell if my blood is in the trees or if the trees are in my blood. That's why it was up to me to find out why the seedlings were fading into corpses.

2

HAND FALLERS

We think of science as a process of steadily moving forward, with facts dropping into place along a neat pathway. But the mystery of my little dying seedlings required me to tumble backward, because I kept thinking of how my family had logged trees for generations and yet seedlings had always taken root.

Every summer we vacationed on a houseboat on Mabel Lake in the Monashee Mountain Range of south-central British Columbia. Mabel Lake was surrounded by lush stands of centuries-old western red cedar and hemlock, white pine and Douglas fir. Simard Mountain, rising about a thousand meters (three thousand feet) above the lake, was named after my Québécois great-grandparents, Napoleon and Maria, and their children Henry—my grandfather—and his brothers Wilfred and Adélard and six other siblings.

One summer morning, Grampa Henry and his son, my uncle Jack, arrived in their riverboat as the sun was rising over the mountain, and we scrambled from our beds. Uncle Wilfred was in his own houseboat nearby. I shoved Kelly when Mum wasn't looking, and he tried to trip me, but we kept it quiet because she didn't like us fighting. My mother's name was Ellen June, but she went by June—and she loved the early mornings on vacation. It was the only time I remember her being completely relaxed, but today we were startled by a howling that drove us over the gangplank bridging our wharf and the shore. Kelly's pajamas had prints of cowboys, and Robyn's and mine had pink and yellow flowers.

Uncle Wilfred's beagle, Jiggs, had fallen into the outhouse.

Grampa grabbed a shovel and bellowed, *"Tabernac!"* Dad followed with a spade, and Uncle Wilfred came racing along the beach. All of us hurried up the trail.

Uncle Wilfred flung open the door, and flies sailed out along with the stench. Mum broke into laughter, and Kelly shouted, "Jiggs fell in the outhouse! Jiggs fell in the outhouse!" over and over, too excited to stop. I crowded in with the men and peered through the wooden hole. Jiggs was paddling in the slop, baying louder when he saw us, too far down in the pit to be reached through the narrow hole. The men would need to dig next to the outhouse, widening the pit underneath, enlarging it until they could reach him. Uncle Jack, half his fingers missing from chain saw accidents, joined the rescue operation with a pickaxe. Kelly, Robyn, and I moved to the sidelines with Mum, all of us giggling.

I ran up a trail to collect a piece of humus from the base of a white-barked birch tree. The humus there was sweetest because this luxuri-

Left to right, brothers Wilfred and Henry Simard with a string of fish at the Simard farm near Huppel, British Columbia, ca. 1920. Sockeye salmon spawned in the Shuswap River and was a main food source for the Splatsin Nation and, later, the settlers. The Simard family logged the forest on the homestead to create farm pasture for raising cows and pigs. When the men lit the slash to clear the ground, the fire escaped up the mountain and burned the forest all the way to Kingfisher Creek, fifteen kilometers (nine miles) away.

Kelly, four, and me, six years old, on Grampa Henry's houseboat
the day Jiggs fell into the outhouse, 1966

ous broadleaf tree exuded sugary sap and shed copious nutrient-rich leaves each fall. The birch litter also attracted worms, which mixed the humus with the underlying mineral soil, but I didn't mind. The more worms, the richer and tastier the humus, and I'd been an enthusiastic dirt eater from the moment I could crawl.

Mum had to deworm me regularly.

Before any ground was broken, Grampa cleared away the mush-rooms. Boletes, *Amanitas,* morels. He placed the most precious—the orangey-yellow funnel-shaped chanterelles—under a birch for safe-keeping. Their apricot aroma stood out even over the drifts from the outhouse. He picked the honey-brown flat-capped *Armillarias,* which were centered in icing-sugar halos of spores. These were not good eating, but the cascade of them around the white-barked birches told him the roots might be soft and easy to break through.

The men started digging, raking the leaves, twigs, cones, and feath-ers into a pile. This sweep revealed an underlying congealed mat of partially decomposed needles, buds, and fine roots. Obscuring these dismembered pieces of forest were brilliant-yellow and snow-white fungal threads coating the collage of detritus, almost like the gauze covering my scraped knee. Through the pores in this fibrous quilt crawled snails and springtails, spiders and ants. To get to the earth's

Moving the Simard houseboat on Mabel Lake, 1925. Grampa Henry and Uncle Wilfred built the houseboat as well as the tug and barge for hauling the horses, trucks, and logging gear to the camps. When there was a good day in the fall, right before the lake froze up, the brothers moved the log booms to the Shuswap River mouth so they'd be ready for the river drive at spring breakup. Uncle Wilfred famously once said, "Only fools and newcomers try to predict the weather."

innards, Uncle Jack sliced his pickaxe through the fermenting layer as thick as the axe-head was wide. Underneath this carpet gleamed the humus, so thoroughly decomposed that it looked like the paste of dark cocoa, sugar, and cream that Mum mixed to make us hot chocolate. I chewed my birch loam intently. Funny, but neither of my siblings or either of my parents ever teased me for my dirt eating. Mum said she was taking Robyn and Kelly back for pancakes, but I wouldn't miss this drama for the world. As the men unveiled another layer, centipedes and sowbugs wriggled through the porous clods getting tossed to the side.

"*Sacrébleu!*" Grampa cursed. The fine roots in the humus layer had become as dense as a hay bale. But he was the toughest person I'd ever known. Once while cutting down a cedar with a chain saw, felling trees on his own, a branch sliced one of his ears clean off. He wrapped his shirt around his head to stem the bleeding, searched for his ear under the boughs, found it, and drove thirty kilometers home. Dad and Uncle Jack took him to the hospital, where the doctor spent an hour sewing the ear back on.

Jiggs was reduced to whimpers. Grampa grabbed a pickaxe and chopped into the cake of rhizomes. The roots were almost impenetrable, forming an interwoven basket of earthen tones. Muted shades of white, gray, brown, and black. A warm palette of umber and ochre.

I savored my sweet-chocolate humus as the men carved into the underworld.

Uncle Jack and Dad made their way through the humus layer and started into the mineral soil. By now the entire forest floor—the litter layer, and then the fermented and the humus ones—had been cleared from an area two shovel blades wide next to the outhouse. A thin bleached layer of sand gleamed, so white it looked like snow. I would later learn that most soils in this mountain country had surface layers like this, as if drained of all life by heavy, percolating rains. Maybe beach sand is so pale because storms quench it of the blood of bugs and the guts of fungi. Among these blanched mineral grains, an army of roots was threaded with an even denser thicket of fungi, sapping the upper soil horizon of any other nutrients that might remain.

Another shovel's length deeper, and the white horizon gave way to a crimson layer. A breeze from the lake rushed over us. The earth had been opened wide, and I chewed my sweet humus faster, like an old piece of gum. It was as though the pulsing arteries of the soil had been revealed, and I was the first witness. I inched closer to see the details of the new layer, mesmerized. The grains were the color of oxidized iron coated in black grease. They looked made of blood. These new clods of soil looked like whole hearts.

The going got tougher. Roots the size of my dad's forearm jutted in all directions, and he hacked them with his shovel. He glanced at me and grinned at the futility of his skinny arms, which made me laugh because we teased him with the nickname "Pinny Pete." Every root looked tenacious in its unique way, though their common job was to graft the trees to the earth. White papery birch, purple-red cedar, reddish-brown fir, black-brown hemlock. Keeping the mammoths from tipping over. Tapping the water that ran deep. Creating pores for water to trickle through and bugs to crawl along. Allowing roots to grow downward to access minerals. To keep the outhouse hole from caving in. Making it hard as hell to dig through.

Shovels were thrown down in favor of the axes to chop through the woody foundation of the forest. Then the spades were used again, only to encounter boulders speckled white and black. Rocks of all sizes, some as big as basketballs and others small as baseballs, wedged into the earth like bricks cemented into a wall. Dad ran to the boathouse for a crowbar. Taking turns, the men levered each rock out of its tight nook—turning, scraping, coaxing. It dawned on me that the gritty soil was a pile of pulverized rock grains. Beaten down by fall rains, dried to dust in summer. Frozen and cracked in winter, then thawed in spring. Eroded by water trickling over millions of years.

Jiggs was buried in a layer cake—the top layer made of fallen plant parts and the bottom of ground-up rock. A meter farther, and the crimson minerals faded to yellow. The colors lightened with depth as gradually as the morning sky changed over Mabel Lake. The roots became sparser and the rocks more numerous. Halfway down the pit, the rocks and soil looked pastel gray. Jiggs sounded tired and thirsty.

"It's okay, Jiggs," I yelled down to him. "You're almost free!"

Grannie Martha had buckets scattered around her houseboat to catch rainwater for drinking, and I ran to get a filled one. I tied a rope to the handle and lowered it to where Jiggs could rest his front paws on it and drink.

It took another hour, and lots of cursing *en français,* before the four men could lie shoulder to shoulder on their stomachs, hanging from their waists into the enlarged hole, to grab his front paws. "One, two, three," they shouted, and Jiggs squealed as they pulled him from the muck. Shuddering, he tiptoed onto a foothold made by that woven carpet of brightly tinted roots and padded over to me, blinking, his orange-black-white fur splotched and clotted with toilet paper. He couldn't even wag his tail. The men were too tired to move, so they pulled out their smokes for a rest. I whispered, "C'mon, boy," and after a few ginger steps we dashed into the lake for a bath.

Later I sat on the shore and threw driftwood into the water for Jiggs to fetch. He had no idea, nor did I, that his adventure had opened up a whole new world for me. One of roots and minerals and rocks that made up the soil. Fungi, bugs, and worms. And water and nutrients and carbon that ran through the soil and streams and trees.

Those summers in the floating camps on Mabel Lake are where I learned the secrets of my ancestors, fathers and sons who spent their lives felling timbers, a history knitted into our bones. The inland rain forests my family had logged seemed indestructible, the big old trees the keepers of the communities. What mattered was that loggers once stopped and carefully gauged and evaluated the character of individual trees to be cut. Transportation by flumes and rivers kept cuttings small and slow, whereas trucks and roads exploded the scale of operations. What was the timber company of the Lillooet Mountains doing so completely wrong?

Dad loved telling Robyn, Kelly, and me stories about his young days in the woods, our eyes big as loonies, especially when the stories were gruesome. Like the time Uncle Wilfred lost his finger in a

Grampa Henry (in white hat), brother Wilfred Simard, and son Odie driving logs through the Skookumchuck Rapids, "the Chucks," at Kingfisher, ca. 1950. The men had to walk, roll, and jump the logs to move them downriver, which was extremely dangerous. Once the logs jammed up in the Chucks, the men had to break them up with dynamite. When he was old and suffering from memory loss, Grampa Henry almost drowned in the Chucks because his outboard motor had quit as he was traveling downriver, and he'd forgotten how to pull the cord to restart the engine. Grannie Martha shouted from the shore until he remembered what to do, just as he was about to hit the rapids.

Fallers on springboards with a crosscut saw at Mabel Lake, ca. 1898.
Two men would take a day or two to fell this western white pine, the most
valuable of the timber species in this mixed forest. Old western white pines
are absent from these forests today because of white pine blister rust,
introduced from Asia at the beginning of the twentieth century.

twisting choker trained around a white pine that Prince, their two-
thousand-pound gray draft horse, was dragging. Grampa stopped
Prince only when Wilfred's screams grew louder than the chain saw.
Or when a cedar pole swooshed onto Grampa's back, leaving him
slightly hunched the rest of his life. They were lucky, in a way; men

Hauling a white pine log at Mabel Lake, ca. 1898. The largest trees in the stand are western white pine and western red cedar, both very valuable for milling into lumber. The large, clear boles and sparse understory show that this primary forest was fully stocked and highly productive.

were routinely crushed by half-downed widow-makers and horse-drawn logs. Some were smashed between jostling timbers, or their hands got blown off by the dynamite they used to break up logjams during the drives on the Shuswap River.

One afternoon the same summer that Jiggs fell into the outhouse, Dad led Robyn, Kelly, and me on a treasure hunt for discarded horse-shoes and chokers along the old flume where he'd worked as a boy. This is where Grampa Henry and Uncle Wilfred had done their hand falling, he told us, bucking (cutting) and limbing the trees. Conifers had been plentiful, with the odd bug or pathogen taking down small clusters of Douglas firs or white pines, or the occasional cedar or hem-lock. The men in my family logged whatever valuable timber they could easily lay their hands on.

Hand falling a single tree took the better part of a day, a week for a patch. Grampa was the jokester next to Uncle Wilfred, a shrewd busi-nessman. Both were inventors: Wilfred built a manual elevator with trolleys in his two-story farmhouse, and Grampa made a waterwheel on Simard Creek to generate electricity for the houseboats. These old forests grew as high as a fifteen-story building, and Grampa would locate the straightest trees. He and Wilfred would stand across from each other on rough-hewn springboards, elevated above the butt swell of the tree where there was a slightly smaller fraction of the girth to cut through. They studied the lean of the tree and the lay of the land, then planned the cuts so the tree would fall in the direction of the flume.

The crosscut saw sang like a slide guitar as the men sweated with each push and pull, and sawdust coated their woolen sleeves as they started with the top cut, slicing horizontally through the trunk on the side of the land's downward slope. A third of the way through the bole—the trunk—they paused for a rest and chewed on smoked-salmon jerky while sap oozed from the cut. Grampa cursed while studying the tree's peculiar lean— *"Il est un bâtard!"*—and pointed his half-chopped-off index finger to warn how the tree could fall in at least two directions. Another hour of aching forearms, and they made a bottom cut at a forty-five-degree angle to the undercut, set to join it deep in the heartwood. *"Mon chou,"* Wilfred exclaimed while knocking the wedge of sapwood out with the back of his axe-head, leaving a yawning grin that resembled their own mouths, since they'd lost most of their teeth to cavities in their teens, now replaced with dentures.

With the face cut completed on the lower side, the men ate strawberry shortcake and drank drums of water. They rolled and shared a smoke. Craven A. Then they climbed back onto the springboards to begin the back cut on the other side of the bole, about an inch above the top cut. Any miscalculation and the timber could buck backward and take their heads off.

They dropped the saw when the tree shifted a tad forward and only a handful of intact fibers were left running up into the heart of the tree. Grampa muttered, *"Sacrament!"* as he pounded a metal wedge into the back cut with the blunt end of his axe. The xylem cracked. With a groan, the tree tilted toward the flume as the hand fallers shouted, *"Timber!"* and ran as fast as they could upslope. The tree whooshed through the air, its crown catching the wind like a sail, creating such an eddy that the ferns below blew forward, revealing their pale undersides. Branches and needles swirling. In seconds, the tree landed with a deafening thud, the ground shuddering. Limbs cracking like bones breaking. A nest of birds catching a draft and floating to the earth in a cloud of feathers.

Grampa Henry and Uncle Wilfred worked along the fallen tree and limbed the branches with axes. They bucked ten-meter-long lengths so that Prince could haul them more easily to the flume. To do this, the men wrapped the end of each cut piece in a choker, as if setting a lasso around a calf—but their "lasso" was an iron chain thick as their wrists. For smaller pieces, they cinched the end of the log using a hand-forged tong that opened as wide as a lion's mouth. They hitched the choker or tong to the whiffletree, a wooden bar carved from a sapling that hung over Prince's tail to tilt and equalize the weight. Prince groaned and snorted as he hauled each bucked log from stump to flume. The brothers then rolled each log into the top of the flume using a peavey, a pole with a swiveling iron hook. Job done, a tree delivered to the water below, they stood sharing another smoke, safe and sound one more day, *one more day*—an image and refrain that still punctuate my images of the hand-falling labors of my family.

I have a tradition of trusting that nature is resilient, that the earth will rebound and come to my rescue even when nature turns violent. But Dad's mother was so keenly aware of the peril in bush work that it gave her pause. She had been crippled in her twenties with drop foot from an infection and wanted the lives of her sons to be freer and safer. Even so, Uncle Jack stayed a logger and worried about his mum to the point of living at home until he was forty.

But Dad left bush work while young. The incident that triggered his decision—he described it for us that day of the treasure hunt, the sun lowering as we sat on logs, the metal chokers we'd happily unearthed piled nearby—came when he was only thirteen and my uncle Jack was fifteen. They'd quit high school to help Grampa Henry and Uncle Wilfred. Their job was to wait on floating logs lashed together with rawhide into a boom on Mabel Lake as each piece of bucked cedar banged the walls of the flume snaking a kilometer down Simard Mountain and thundered toward them like a luge. Once the log hit the water, it was up to Dad and Uncle Jack to steer it into the boom.

While shivering one morning in the spring rains, Dad panicked.

Logs torpedoing down one of Grampa Henry's flumes into Mabel Lake. This flume emptied near the outlet of Simard Creek, where Grampa had also built a waterwheel to generate electricity for the loggers' houseboats.

River drivers on a log boom at Mabel Lake. Wilfred Simard, third from left, is holding a four-meter (fourteen-foot) pike pole used for guiding the timbers. The shorter peaveys ended in a metal U-hook and a spike for helping the men turn the logs and keep their balance. The work was dangerous, but a driver falling off the logs was considered a sissy. The shorter Douglas fir logs in the foreground of the boom were sawn into lumber, and the longer cedar logs at the back of the boom were sold for hydro-telephone poles. The cedar poles brought more profit but were much more difficult to drive because they would jam up in the flowing river.

Wooden pike pole in hand, iron lance spliced to its end, he tried to stay balanced on the log rolling below him. "It's coming!" Jack shouted, his feet barely keeping pace with his own churning log while Dad's gathered momentum as waves lapped. The cedar log launched from the bottom of the flume, like a skier off an Olympic jump, arching higher than usual before piercing the water twenty meters in front of them, straight into the bottomless lake. There was no telling where it would explode back like a missile to the surface.

Time stopped. Dad told us his mind had snapped back to the essay he'd written about World War II before he quit high school, when he wrote, "All night long, the cannons went boom, boom, boom . . ." His teacher had asked for five hundred words, but Dad had no idea how to string so many words together to describe a soldier's terror. He was sure the log would shoot up and pulverize him.

"Run, Pete!" Jack yelled.

But he couldn't, not even when Jack ran toward shore, screaming at Dad to follow him, to get the hell out of any chance of the log's path. Dad couldn't hear a thing. Seconds ticked by.

Boom! The log shot sky-high twenty meters behind him before landing with a swoosh. Tremors ran through Dad's hands as he trained the bobbing log into the boom. In the fall, Grampa's boat *Putput* would tug the boom downriver to sell the largest logs to sawmills, with the smaller-diameter cedars going to Bell Pole Company for telephone poles.

He went into grocery management not long afterward and stayed there his entire career. But the forest would always be our life's blood.

THE TRAILS still remained from those long-ago logs gliding along the forest floor. Perfect landing places for seeds, some as small as a grain of sand, others the size of opals. Seeds of western red cedars and hemlocks were from cones the size of a man's thumbnail. More seeds came from Douglas-fir cones the size of a fist, others from white-pine cones as long as a forearm. In the patch mowed by the dragged trees, the seeds of the old trees had sprouted into a dense sward of seedlings with white-tipped roots tapped into the humus and pools of water. They were tough, their genes shaped for resilience by their elders

Grannie Martha, about twenty years old, log walking on Shuswap River at Kingfisher, ca. 1925

over generations. All the species of the forest were layered according to their growth rates. The prominent Douglas firs and white pines towered over the group in the middle of the gap where the mineral soil had been exposed and the sun had shone the longest, and bendy cedars and hemlocks, already as tall as I was on the afternoon of our treasure hunt, lounged in the shade of their parents. The Douglas-fir saplings in the center of the haul trails were twice as tall as Dad.

The hand falling, horse logging, and river drives left the forests capable of vibrant, renewed life. Clearly much had changed from what I knew to what my industry and I were doing now.

I stared out a window in the Woodlands office and thought about my plantations. There were many ways to improve—sow more locally adapted seed in the nursery, grow bigger seedlings, prepare the ground more meticulously, plant sooner after logging, remove competing brush. But the clues told me the answer lay in the soil and how the seedlings' roots connected to it. I sketched a robust seedling with branched roots and running fungi, and a sickly one with a minuscule shoot and stunted roots. But my ideas would have to wait, because today I was assigned to work with Ray in a two-hundred-year-old forest in the glaciated Boulder Creek valley, a couple dozen kilometers from Lillooet.

On this day, I was to play the role of executioner.

Ray and I were there to mark a boundary for clear-cutting. He wasn't much older than I was, and he lived with the rest of us students in the bunkhouse, but he had experience working in the steep Pacific coastal terrain, and he reminded me of the men in my family. He'd already lost a piece of flesh in the woods, mauled by a grizzly who'd lifted him by the butt with her teeth and was carrying him away before his compass man scared her off with a shotgun.

We headed past the grinding excavators and scraping graders that were building a new haul road and stopped near some old trees on the loamy fans that had accumulated in the crease of the valley. Engelmann spruce, with its broad sweeping crowns and mammoth gray boles. Ray had flashed the map at me—he wasn't used to sharing information with a girl, and he'd been in a hurry—but the contours I'd glimpsed showed the slopes climbing to towering ridges, the forest increasingly sparse as it met the rocky talus where the marmots perched. Spruce along the creek gave way to Douglas firs where the soil pockets were deep enough to support a sprawling root system. The forest was interrupted every few hundred meters by sweeping avalanche tracks, where devil's club as thorny as roses and lady fern as lacy as petit point were waist-deep. I remembered these same plants at Mabel Lake. Elation spun in my chest, but it was pinned below the lump in my throat. I picked a sprig of foam flower, its tiny white flowers like ocean spray.

With his red wax pencil and compass, Ray marked a perfect box on the aerial photograph where the timber would be clear-cut. He rolled the photo and wrapped an elastic band around it.

"Oops, Ray, missed it," I said. "Could you show me again?"

He reluctantly produced the map, his expression unreadable.

"Are we going to take it all?" I asked. "Couldn't we leave some of the oldest ones?" I pointed to a monumental tree with lichens drooping like curtains from its branches.

"Are you an environmentalist?" He was a precise technician, aligned with the times and the job. This was his profession, he loved it, and he was getting paid to get it as right as he could.

I looked at the dead-forest-standing. I was thrilled to be working in this venerable expanse; I didn't even mind figuring out how to log some trees. But obliterating whole tracts in one fell swoop would leave little foundation to help the forest recover. The trees grew in clusters, with the eldest and largest—one meter in girth, thirty meters in height—in the deepest part of the hollows where the water collected, with younger trees of various ages and sizes close by. Like chicks clutched around a mother ptarmigan. The grooves of their bark housed tufts of wolf lichen, easy for the deer to nibble in winter. Buffaloberry and soapberry shrubs grew between rocks. Bright-red Indian paintbrushes, purple silky lupines, pale-pink Calypso fairy slippers, and candy-striped coralroot traced the roots fanning out from the tree boles. None of these herbs would thrive after a clear-cut. What the hell was I doing here?

Using Ray's calculations, we marked out the square with pink ribbons hung every ten meters or so. The fallers would see the pink edge and know where to stop clear-cutting. The ancient ones outside the boundary would be spared.

Ray told me to run the line straight at 260 degrees, almost due east, basically following the edge of the avalanche track. He was gazing up the boundary as I pulled the chain, a slippery nylon rope, fifty meters of wrapped twine, from the back pocket of my vest. He would follow, filling in with more flagging for the fallers.

I adjusted the dial of my compass and spotted a tree to use as a beacon. The chain unraveled like a skipping rope, every one of its fifty metal clasps marking one-meter increments. I moved like a coyote, threading the chain over logs, through thickets of brush and between families of trees.

"Chain!" Ray shouted when I reached the end of the fifty-meter stretch. When he tugged on his end of the rope, I hung a ribbon to mark the spot.

"Mark!" I shouted back, my voice rising over the sound of water rushing below. I loved shouting, "Mark."

Satisfied with the accuracy of our first chain length, Ray climbed toward where I was hanging pink ribbon from branches. A squirrel jabbered from his perch, and I dug my fingers into where he'd excavated and I felt a soft pebble. Under the forest floor was nestled a piece of fungus like a chocolate truffle that I pried out with my knife, cutting a black strand that plunged deeper into the soil. I stuffed the truffle into my pocket.

"You see those big pumpkins?" Ray asked, referring to some large firs outside the line of our square. He thought we should take them. The bosses would be pleased—an extra bonus of prize trees.

I pointed out that they were well away from the cutting-permit borders. Including them would be illegal. Not only were big elder trees an important seed source for the open ground, they were favorite perches for birds, and I'd seen bear dens under the necks of the roots.

Neither of us had the authority for decisions like this. I knew he loved the trees too; that was the fundamental reason we both picked this profession. "Can't leave perfectly good firs for no reason," he said, mulling it over. "They can be run through the veneer mill."

We walked to one of those forbidden elders, and I wanted to shout at it to run. I understood the pride of claiming what was grandest, the temptation—green-gold fever. The handsomest trees captured top prices. They meant jobs for the locals, mills staying open. I checked out this one's immense bole, seeing the cut through Ray's eyes. Once you start hunting, it's easy to get addicted. Like always wanting to snag the tallest peaks. After a while, your appetite can never be sated.

"We'll get caught," I argued.

"How?" Ray's arms were crossed, his face puzzled. The government couldn't check every inch of our block boundary. Besides, these ones were so close, so easy.

"They're habitats for owls." I had heard at school about the rare dry-forest ones—flammulated owls—but didn't know much about them. I didn't have a clue if they were at Boulder Creek. I was grasping at straws.

"Do you want this job next summer? I sure do." The company would give us credit for finding more wood. He glanced backward, as if the tree might pick up and escape.

I wanted to scream at the top of my lungs. Instead I rerouted the line and cried inside at my weakness. At the timberline where a magnificent fir stood, my shoulders tightened. A curtain of cow parsnip and willows obscured the avalanche track, but the air was still. I quickly hung the pink ribbon so the tree fell inside the boundary. In a week, it would be lifeless. Delimbed, bucked, and piled along a road right-of-way, waiting to be loaded onto a truck.

Ray and I rerouted all the borderlines. We condemned another ancient.

And another. And another. By the time we were done, we'd stolen at least a dozen elders from the edges of the avalanche tracks. When we took our break, he offered me chocolate chip cookies and said he had made them himself. I declined and wrapped the nylon chain into a figure eight, using my boot and knee as anchors. I floated the suggestion that we could convince the company to leave some firs in the middle of the block to spread their seed. I blurted, "You know, the way they sometimes leave the big seed trees in Germany."

"We only clear-cut around here."

When I tried to explain that where I'd grown up, we'd log small patches, and the dragged logs churned the floor to make a bed for fir seeds to germinate, Ray countered that if we left some lone firs, the wind would blow them over, and the bark beetles would move in. "And the company will lose a pile of money," he added, frustrated that I just didn't get it.

It would be a gut punch to see the stately firs reduced to stumps, the elegant shape of the stand carved into an empty square. Back at the office, I glumly prescribed cluster plantings for the clear-cut, with Douglas fir in the hollows, ponderosa pines on the outcrops, and prickly needled spruce along the creek, mimicking the natural patterns. Ray was right, of course, that the company would reject my idea of retaining a few elders for seeding the disturbed ground, but this planting design would at least maintain the natural species richness of the site.

Ted told me we would just plant pine.

"But there's no lodgepole pine up there," I said.

"Doesn't matter. It'll grow faster, and it's cheaper."

The other summer students near the map table shifted. Foresters in the surrounding offices put their hands over their phone receivers, alert to whether I would have the guts to start an argument. A calendar fell off the wall, struck the floor.

I went to my desk and rewrote the planting prescription, and my heart withered. What had happened to that little girl who ate dirt? Who'd made braids of roots, entranced by complex natural wonders? Places of terrible beauty and layered earth and buried secrets. My childhood was shouting at me: *The forest is an integrated whole.*

3

PARCHED

I stood straddling my bicycle and took a long drink of water. It was high noon, and the sun was beating down on the dry forest. I'd ridden a hundred kilometers, and the heat was sucking the sweat out of my summer-dark skin. The low-lying mountains of southern interior British Columbia stayed bone-dry because the east-flowing Pacific air dumped most of its rain on the coastal mountains, which stretched two hundred kilometers from the ocean, to twenty kilometers west of here, leaving this interior blue sky with nary a drop. This weekend, I felt pure freedom in this landscape, the tension with Ray over the ancient Douglas firs out of my thoughts and my disappointment with Ted's decision about the planting prescription buried.

I was on my way to see my brother Kelly compete in a rodeo, among the cowboys and horses where he belonged. I'd last seen him at Mum's a couple of months back and caught him sobbing because his girlfriend, a barrel racer, had left him for another man while he was at farrier school in Alberta. We'd stood in the dark, him leaning against his brass-colored truck, the bed holding his new horseshoe forge and anvil. His head was bowed as he tried to swallow his grief, but he couldn't, and I cried with him.

I peered down the valley a handful of kilometers to where a river ran through a hollow cloaked in sage and grass. Those rugged, knee-deep perennials were the only plants that could find footing in that arid soil. Trees needed too much water to survive down there. But up here there was just enough water for trees to establish in recesses among the grasses, forming open woodland.

An afternoon haze was developing, likely from a wildfire, but it was still clear enough to see the valley rise a thousand meters to the next ridge, yet another half-dozen kilometers away. As the elevation increased, so too did the rainfall, and the branching gullies soon filled with curvy lines of trees tracing the flow of water. The trees up the gullies eventually spilled onto the knolls, and the forest filled in to form a continuous cover. As the mountain forests rose further still, the trees again grouped on hummocks to escape the cold, wet soil, until they dwindled altogether to be replaced by pale green alpine meadows.

I dropped my bike and took a short walk into the grassy woodland for some shade, through patches of Douglas firs and under parasols of ponderosa pines in depressions where trickles of water collected. I scrambled onto a knoll where a single ponderosa pine grew, its long needles in scanty bundles to save precious water. This afforded ponderosa the distinction as the most drought-tolerant of all the tree species in these parts. This one was in an especially precarious position, where even the deep-rooted bunchgrasses had turned brown and shriveled to minimize water loss. I turned my water bottle upside down to give the pine the last drops and laughed at my gesture. Only its taproot could save it in times like this.

A grove of old Douglas firs occupied a shallow gully, and I beelined toward it. Puffball mushrooms blew clouds of brown spores at my face; grasshoppers clicked their legs. Kelly and I used to collect the mushrooms to make puffball soup, and I picked one, fungal threads flowing from its fulcrum. I thought I'd give it to him. He'd love that I'd found it on the grassland, because foraging was among our favorite childhood games.

The crowns of the elder firs cast ample shade. They grew in these draws because their dense bottlebrush needles required lots of water, at least compared with the sparsely needled ponderosa pines. This restricted where they could grow but also enabled them to become taller and form denser clusters than the pines. But Douglas fir and ponderosa pine were both better than the spruce and subalpine fir at minimizing water loss, helping them cope with the drought. They did this by opening their stomata for only a few hours in the morning

Taking a break, age twenty-two, under a Douglas fir between Enderby and Salmon Arm, 1982. My friend Jean and I spent many weekends in the early 1980s touring the interior roads with only our sleeping bags and ten bucks in our pockets. I had lost my wallet that day, and when I got home a motorist had called Dad to say he had found it on the side of the highway with my license and ten dollars tucked inside.

when the dew was heavy. In these early hours, trees sucked carbon dioxide in through the open pores to make sugar, and in the process, transpired water brought up from the roots. By noon, they slammed their stomata closed, shutting down photosynthesis and transpiration for the day.

I sat eating an apple under the generous crown of an old Douglas fir, the seedlings on the outskirts of its apron a sign that the ground was cool and moist. The brown furrowed bark absorbed the heat and protected the tree from fire. It was thick, too, to prevent water loss from the underlying tissue, the *phloem,* which transported the photosynthetic sugar water from the needles to roots in an inch-thick ring of long tubular cells. The orange bark of the ponderosas also protected the parasol-crowned trees from the fires that swept through every twenty years or so.

These seedlings were growing happily where there was barely any water, while my seedlings in the Coast Mountains to the west were dying where there was plenty.

The awn of a renegade grass seed head tickled my bare leg as I glanced at an ant crawling over from a nearby nest as tall and wide as my seated silhouette. The nest quivered with thousands of workers. Moving, stacking, and stockpiling millions of Douglas-fir needles that littered the forest floor. The ants also carried spores of brown decay mushrooms on their legs and in their fecal pellets into the nest, accelerating infection and decomposition of the needles, which settled and stabilized the thatch. And into stumps and fallen trees, aiding decay otherwise hindered by the summer drought. I remembered the saprotrophic oyster mushrooms at Mabel Lake, their smooth, creamy caps attached to the fallen leaves and logs of dead birch trees. Trees that had been killed by the pathogenic honey mushrooms. The oyster mushrooms, their decay skills so efficient, also killed and digested bugs to meet their needs for protein. Mushrooms were as varied as their roosts, and they were masters at multitasking.

Somehow, in the ravines and hollows of this parched valley, the saplings and seedlings sprinkled around the Douglas firs and ponderosa pines seemed fine—without the benefit of a deep taproot of their own yet. Could the old trees be helping the young ones by passing them water through root grafts? Grafts were unions where roots of different trees spliced into a single root, with phloem shared in common, like veins grown together in a healing skin graft.

It was time to go or I'd miss Kelly's bull-riding event. He rode bulls because it was the cheapest event to enter, and he was always broke.

Still puzzling over the water enigma, I headed back to my bike and noticed a cluster of aspens across the road, their bark smooth white. These too had spread up from the wetter gullies onto rockier slopes. They had big flat quivering leaves that surely emitted gallons of water every day. Trembling aspen are unique in that many stems of the same individual spring from subterranean buds along a shared network of roots, and I wondered if the aspen copses were accessing water from the ravines and passing it upslope through their shared root systems. Like a fireman's brigade. Under their crowns, wild roses sprouted, pale pink petals wide open to flaunt bright yellow stamens. Kelly's favorite flower. Knots of purple silky lupines, golden heart-leaved arnicas, and rosy pussytoes spread from the shade into the sun. Was the root system of the aspen leaking some water into the soil for them to access? Maybe this was how the riotous plant community survived in the shallower, drier soil. But I had no clue how the water got from the old aspen trees to the little flowers without first evaporating in the sun.

I stopped at a tortuous ponderosa and dug a hole in the lichen-crusted soil to bury my apple core. The hard clay was matted with tree roots and grass rhizomes—the underground creeping stems with nodes here and there, like the runners of a strawberry plant. Though dry, the mineral clods were chock-full of voluminous white and pink and black fans of fungal threads. Thinner than the fleshy strands I'd seen in childhood when Jiggs fell into the pit of colorful roots and soil. Finer than the thick yellow mats in the subalpine-fir forest below the clear-cut earlier in the spring. A pink coral fungus, so named because it looked like coral on the ocean floor, poked out of a bed of lichens forming a crust over the ground. I picked the tiny fungal tree, only an inch tall, to take a closer look at its delicate upright branches. They were clearly as efficient as the gills, pores, and ridges of the other fungal species at creating ample space for spore production—millions blew up my nose, and I sneezed. Pink fungal fibers fluttered from its base.

What were the fungal threads of this odd-shaped mushroom doing, and how were they helping the coral fungus make a living? I rubbed the threads between my thumb and index finger. They were gritty. Moist soil particles clung to the mycelia. The threads might have a role in gathering water from the labyrinth of pores in the soil. In this climate, any water still in the ground would adhere to soil particles with the strength of cement. In the sparse woodlands, where trees only grew in depressions and gullies, water was obviously limiting where they could get a toehold. I wondered if these tiny mushrooms might be helping not only themselves but the trees in need of water or perhaps nutrients where the trees were surviving the cold. If I rode my bike to those high-elevation forests across the valley, would I find the *Suillus* pancake mushrooms there too, as I had in the Lillooet Mountains? In places with plenty of water, maybe the pink and yellow and white threads were delivering nutrients instead of moisture to the trees. I put the coral fungus into my pocket with the puffball.

More puzzling was the question of whether the multitude of silky fungal threads fanning through the clay could explain how water moved from the big trees to the more shallow-rooted plants. Were those threads, which looked like an underground spider's web, joining trees and plants together to capture much-needed moisture for the whole community? Were the puffball and coral fungus involved? Maybe they had nothing to do with it, since the prevailing wisdom was that trees only compete with one another to survive. That's what forestry school had taught me, and it was why my logging company liked fast-growing trees spaced well apart in rows. But that didn't make sense in this ecosystem where trees and plants seemed to need one another for survival. One extremely dry season, a profound dryness the trees were not adapted to cope with, and they could succumb to the blistering heat.

IN THE NICK OF TIME AS USUAL, I arrived at the Logan Lake arena as Kelly's event was starting. The rodeo grounds were in the center of the village, which was nestled in the low glaciated interior range covered with pale, arid fir and pine forest and grassy meadows. Only a few thousand people lived here—ranchers, loggers, copper miners. The unobtrusive mountains, formed of compacted till and volcanic pistons, weathered over millions of years, reminded me of the solid, hardworking people they surrounded. The sun bore down on the dusty grounds, warming the earth and intensifying the aroma of the horses and bulls. Dogs drank deeply from water bowls in the shade, and kids hung out under the awning over the fishpond. Cowboys and cowgirls led their stunning rides—Appaloosas, quarter horses, paints—between the stable and ring. The crowd was settling in for the bull riding as I found a place low in the grandstand and scanned the chutes for Kelly's brown-felt cowboy hat.

The cowboys were, in spite of the heat, in full regalia, with yoked embroidered western shirts and tight creased jeans, as elegant as Elizabethan aristocracy. I pulled my ball cap low over my eyes to block the sun, wishing I had a Stetson. My T-shirt and shorts were insufficient. In these low-lying mountains, it was hotter than the hubs of hell, and exposed skin burned in minutes.

Then I saw Kelly.

He was straddling the fence surrounding the competition chute holding his bull. The chute, barely wider than the bull itself, was at the far end of the oval arena, closed off with a gate. A clown was inside the ring. Kelly's legs were taut under jeans and leather chaps as he waited for the bull to calm down a little. With a grin, he talked to the animal. His clear blue eyes were so focused they looked anchored under his dark eyebrows, and his worn leather gloves enlarged his already huge hands. I knew his leather belt was inscribed "Kelly," cinched with a silver trophy buckle engraved with a mountain lion, fitting testament to this cougar country where we'd been raised. Where our parents had taught us how to camp. How to make a garden and catch a fish. How to paddle a canoe to the stockyard so we could ride Kelly's horse Mieko. Where together we'd learned our place, our meaning, our reason in the wild. Building tree forts, battling in shoot-'em-ups. Making swings of long ropes and rickety rafts in the cool rains of Mabel Lake. As a kid, Kelly had practiced for hours on the blue barrel strung between cottonwoods. Robyn and I would use our full weight to work the ropes while he rode the barrel like a bucking bull with his imaginary spurs.

He had drawn the meanest bull—Dante's Inferno. The scoreboard flashed Dante's statistics: he'd bucked off 98 percent of the cowboys who'd tried to ride him, and he'd scored 45 percent for his spins, kicks, drops, and rolls. Fifty points were assigned to the bull and fifty to the cowboy for how fluidly he matched and countered the bull's moves. Kelly waited on the fence as Dante crashed against the chute's walls. Cowboys in the stand shouted hoarsely. The clown danced, ready to open the gate wide. Kelly looked up and searched the crowd. Drawing Dante was a double-edged sword. Getting bucked off before the full eight agonizing seconds meant no score, but staying on could yield more points for the quality of his ride.

Frothy saliva streaked Dante's hide, his frustration at being trapped amplified by the crowd. I pictured the scar below Kelly's lower lip, stretched by the familiar lump of chaw tucked against his gum. He got that scar when he was eleven from smashing his bike into a parked truck because we were racing to see how high my new speedometer would go.

He saw me in the stands and flashed a smile. *Don't worry. I've got this.*

I nervously rolled the coral fungus between my fingers.

The announcer chattered over the loudspeaker as the bull heaved and bucked. I stiffened with pride as he introduced Kelly as an up-and-coming star. He was already well-known in the little British Columbia towns of Chetwynd, Quesnel, and Clinton for hanging on to his bulls. The winnings were money, what most of these cowboys were well short of. Five hundred dollars for today's champion in this small-time circuit. Kelly joked with the clown by pretending to plug his ears at the bull's raucous banging. The clown had a white-painted face with red lips and wore a yellow-checkered cowboy shirt with baggy jeans.

"Hey, clown," the announcer teased over the loudspeaker.

The clown did a cartwheel. "What?" he shouted.

"Where does a cowboy cook his meals?"

The clown shrugged but kept a practiced eye on the chute.

"On the range."

The crowd shrieked with laughter as the clown fell to the ground to prove how much he was suffering. Kelly was ready on the edge of the chute. The bull was settling a little.

"Hey, clown. You hear about the three-legged dog? He walked into a saloon and had a question for the bartender."

The clown placed both his hands on his hips and shook his head, because dogs can't talk.

"I'm a-lookin fer the man who shot my paw."

The clown banged the heels of his hands on his head, and the crowd howled but then fell instantly silent.

I glimpsed Uncle Wayne, my mother's brother, a few rows in front of me, focused on Kelly as if silently coaching him. Kelly was Wayne's protégé, and Wayne was Kelly's idol, both born cowboys from the ranching line of Fergusons. Hard-living men who would rather die riding their horse across a meadow than in a chair reading a book.

I wasn't made from the same maverick fabric, but I knew bull riding was *the* most important thing for Kelly to do, in his blood as much as the trees were in mine.

Dante suddenly understood his predicament and came to a standstill.

Kelly tipped his hat to the referee, who was sitting on the rail on the opposite side of the chute, wrapped his right wrist tightly with the braided rope looped around the front of the bull's torso, and straddled Dante. The rawhide strips flowing from the wristband of his glove looked graceful against the strength of his arm and the power of the bull. When Kelly nodded, the referee yanked the thick flank strap down hard to tighten it around the bull's groin.

The clown swung open the gate, and the bull roared out, kicking, twisting, and swinging. The crowd rose and screamed. The arena shook. Everyone was electrified about *my* little brother. The flank strap was doing its job, digging in to make the bull buck crazily with its hind legs. A lanky cowboy behind me screamed, "Ride 'im, you son of a bitch!"

Kelly clung to the rope with his right hand while he threw up his left arm. I swallowed my distress. Dante spun, all legs in flight, while Kelly hung on, moving in stunning precision with the bull's kicks. It charged so close to the arena's edge I thought they were going to crash through the boards. The bull bellowed as Kelly's spurs raked his hide. I was schooled enough to know the referee would give Kelly more points for provoking Dante's fury. Every tendon bulged in Kelly's neck. The clown waved his red handkerchief to draw the bull back toward the center.

As the clock ticked toward eight seconds, I threw my fists to the sky and yelled until my throat ached. But I also knew that one unexpected twist, perhaps prompted by a shrill shriek from the audience, and Kelly could be a rack of busted bones.

I glanced away but forced myself to see the violent buck that threw him. He soared, arching high before landing shoulder-first with a horrid thud. The blood rushed from my head. Kelly jumped out of the bull's path in the nick of time. The crowd groaned and sank back into its seats. The countdown clock showed seven seconds. Uncle Wayne shouted, "Geeezuz keerist!"

Lithe as a gymnast, the clown leapt in front of the bull so it would chase him while Kelly staggered to the fence. A cowboy galloped his horse alongside Dante and grabbed the flank strap. The buckle unhitched, and the strap fell into the dirt. Dante kicked out his last buck and charged around the arena, gradually slowing until the cowboy was able to steer him into the adjoining corral.

"Give him a hand, folks!" the announcer bellowed. When he made the customary proclamation of, "He paid his entry fee!"—a show of respect for cowboys who were thrown—the crowd clapped. The next cowboy was already in the competition chute.

Uncle Wayne, a calf roper well loved on the circuit, with a reputation for meticulous cattle ranching, prodigious ranch sales, and hard drinking, was jawing with some cowboys, waving his arms in emulation of Kelly as the men yelled about the minutiae of the seven seconds.

I got to the first-aid trailer, its metal walls boiling hot, as the medic was relocating Kelly's right arm into its socket. His shirt looked clean but was crumpled in a ball. The medic manipulated Kelly's shoulder, and it had to hurt like hell, but he looked happier than a pig in shit. No sign of the anguish over losing his barrel-racing girlfriend. His drooping arm made me want to retch. Several girls entered, their tailored shirts fit into even tighter blue jeans, held up with silver studded belts and tucked into ornately stitched cowboy boots. How had my family missed out on so much strut? A shy girl at the back of the clutch, with raven hair and green jeweled eyes, caught Kelly's attention, and he smiled at her and waved at all his admirers.

The medic gave a final twist, and Kelly suppressed a moan as the ball of his humerus slipped back into the cup of his scapula. The girls, more accustomed to this sort of pain than I was, since they were ranching people themselves, crowded closer in awe. But my stomach was swirling, and I stepped to the doorway.

Overwhelmed by the attention, Kelly called over to me, "Hey, Suzie, you ride all this way in the heat?" He was grinning. The raven-haired girl must have sensed I was his sister, because she dropped back a bit, ceding time and space for me to be with him while the other girls drifted away.

"Yeah, but I started early." I propped myself next to him on the wooden medical table.

"Second time I've done this. The doc says it'll be easier to dislocate each time."

"You'll recover." I didn't want him to have to quit. He was hitting his stride. I hadn't seen him this vivid, this alive, since we were kids.

Kelly laughed, and in spite of the pain, flexed his left arm to prove I was right. "You look pretty skookum, too," he said.

It felt good to have a normal conversation. When our parents' marriage had ripped apart, Kelly fared much worse than I did. He was younger and the only one of us still living at home when they each landed in the hospital, unable to cope. Mum tried to reassure me she'd be fine when I visited her in the ward, but her confusion over why she'd ended up there didn't convince me she was improving. In his apartment after his own hospital stay, Dad dragged on smokes and stared at the walls. I wanted to scream at them to get their shit together, but mostly I wanted to cry. Kelly moved from Mum's house to Dad's, and back again, before and after their recoveries, desperate for enough stability to finish high school. He took Dad fishing, Mum

Kelly in his midtwenties bull riding at the Falkland Stampede, ca. late 1980s

skiing, but he couldn't break through their sadness. He'd explode with frustration, yelling over nothing. I accidentally honked my horn once while he was working on his truck, and he charged out of the garage, shouting at me. Meanwhile Robyn had jumped around in her studies at the university and had taken a year off to travel. We tried to find solace in one another, but as young adults without a home to return to, we scattered.

But on the rodeo grounds, being with Kelly felt like the old days in the woods when we were building camps and riding the trails.

The raven-haired girl was patiently standing by, and Kelly asked her name. Before she could respond, the trailer shook as Uncle Wayne burst in and exclaimed, "You drew the worst goddamned bull of the rodeo!" His trophy buckle, big as a dinner plate, boasted a longhorn steer.

"Yeah, the bastard was crazier than a shithouse rat," Kelly replied, spitting a stream of tobacco into a spittoon. "Gave me a run for my money."

"Wasn't that bull somethin' else, Susan?" Uncle Wayne boomed. He always got my name wrong. I nodded, agreeing. Wayne looked at the girl and said, "Hi, Shen. Your calf roping is coming up fast, can't wait to see you ride. How's yer dad? Still working at One-Fifty Mile House?" This was a juncture on an old gold-rush road, a stopover point with a store and gas pumps.

"He's fine," she replied, appearing amazed that he knew about her family. Uncle Wayne made it his business to know everything about everyone.

"I had a friend who lived at Lac La Hache, not far from the One-Fifty," I offered, not knowing what else to say.

When another girl wandered in and offered Kelly aspirin, Shen floated to the exit, and Kelly watched her disappear. As far as I knew, he never saw her again, but I'll always be grateful for what she gave him that day: her unabashed respect, approval, fondness. Her sudden wish to flee wasn't different from my own impulses. Kelly understood when people like me had to vanish into thin air, just as I knew when he was feeling buried by rapid changes, as if he'd been born a century too late. I thought about showing Kelly the puffball but didn't want to

embarrass him in front of Uncle Wayne, so I threw a jab at his unin-
jured bicep to wish him goodbye.

"Hey," he said, "thanks for cycling all this way in the godforsaken
heat just to see me."

"Anytime," I said, laughing back. "Where's your next rodeo? Maybe
I can make it."

"Omak, Wenatchee, and Pullman," he said. "All in one weekend."

"Geez," I replied. "Outta my league. Good luck. I'll catch you next
time you're around." We had run out of words, even though there was
still so much to say.

Kelly tipped his hat at me and inserted a new hunk of tobacco
under his lip.

I SPED ON MY BIKE through the Douglas-fir forest to get back to
my Volkswagen Beetle, which ran pretty well as long as the gearshift
stayed wired in place with a coat hanger. I was due at the Woodlands
office early in the morning and regretted being shy about asking Kelly
what he thought of my seedling enigma. He would have thought long
and deliberately, then come up with an answer I'd never considered.
Like the time he'd braided cottonwood whips to fix my busted reins
when we were riding horses. I could find the good strawberry patches
in the pine flats near our house; he could deliver calves and cauterize
wounds on the range. He worked through problems by comprehend-
ing the basic order of things and offering something brilliant. Explaining it in a few words. Followed by a laugh, then stillness.

Halfway to my car, it occurred to me that I was starving, and I
stopped under a Douglas fir to eat my cheese sandwich while a squirrel chattered at me. He held a chocolate-brown truffle coated in a
black rind, gnawing at it with the cadence of a hummingbird. He had
excavated it from the soil of the fir. Several burrows were lined with
piles of fresh soil from his exhumations.

"I'm not sharing," I said. "You've got a truffle." I ate in a hurry and
grabbed my knife from my pannier, shooing the squirrel so I could dig
around one of his holes. He moved to his midden and prattled loudly,
still chewing his truffle, its spores flying.

I dug through sheets of hard clay, each layer cloaked in black fans of fungal threads. I held a clump close to my eye and saw the tiny threads growing straight into the soil pores. Working my knife through the layers, I realized that every single sheet was coated in the fungal network. I hit a soft spot, as if I'd poked a cooked potato, and carved through the clay until a dark, round truffle stared back at me, its black rind fissured. I swept aside the surrounding soil, as though I were on an archeological dig looking for shards of bone, until I could get my fingers all the way around the tuber.

As the hole grew to the size of my foot, I uncovered a fungal strand flowing out of the truffle. It looked like a thick black umbilical cord, wiry and tough, and made of many individual fungal strands twisted and packed together like ribbons around a maypole. The strands themselves came out of the black fans caking the clay sheets before knitting themselves into one. The cord was packed into the clay, so I chiseled off more soil to see where it went. With about fifteen minutes of work, I followed it to a whitish-purplish cluster of fat Douglas-fir root tips. I poked the tips with my knife—they had the same softness and texture as a mushroom.

I stared at the excavation, my mind whirling. The cord was linking the fungal-coated root tips of the Douglas fir to the truffle. The tips were also the source of the fungal threads fanning over the soil pores.

The truffle, the cord, the hyphal fans, and the root tips were tethered into a single whole.

The fungus was growing on the roots of this healthy tree. Not only that, it had sprouted a belowground mushroom, a truffle. The relationship between the tree and the fungus was so robust that the fungus had borne fruit.

Exhaling, I rocked onto my heels. Since the root tips were coated in the fungus, any water the roots would be accessing, or anything soluble in water for that matter, such as nutrients, would have to filter through the fungus, which looked as if it had all the tools to act as the joiner between the roots and the soil's water. Out of the fungus flowed a whole underground apparatus—truffles, cords, and strands that in turn grew fans of ultrafine hyphae that infiltrated the soil pores. These pores were where water was held so tightly that it would take a million of the microscopic threads to suck up enough to make a drop. The fans could be soaking up the water from the soil pores, then funneling it to the strands that formed the cord, which then passed it to the attached fir root.

But why would the fungus give up its water to the tree roots? Maybe the tree was so parched, with such a deficit from transpiring water through its open stomata, that its roots sucked the water from the fungus like a vacuum cleaner. Or a thirsty kid drinking through a straw. This exquisite underground mushroom system sure looked like the lifeline between the tree and the precious water in the soil.

After half an hour as an impromptu archeologist, I needed to skedaddle. I wrapped the truffle, cord, and attached root tips in the wax paper from my sandwich, packed my treasure into my weathered red pannier, hopped on my bike, and waved goodbye to the squirrel, still feasting on truffles. Pedaling hard, I reached my Volkswagen Beetle as dusk was falling, strapped my bike on the top with rope, and pulled on a sweatshirt. With one bicycle wheel hanging over the front and the other over the back, my old blue Bug looked as if it had sprouted butterfly wings.

I wound along the Fraser River to Lillooet, so tired I began nodding off, snapping awake to imaginary deer running onto the road, arriving at the company bunkhouse before midnight. I tiptoed down the hallway past the cramped bedrooms where the four other summer students, all young men, were asleep. In my shotgun bedroom— it felt like a walk-in closet—I looked for the mushroom book from the library. My room was a mess, and I wished that I'd inherited my father's fastidiousness. *Aha.* The book was under a pile of jeans and T-shirts.

I paged through it. The puffball was the species *Pisolithus,* and the coral fungus was *Clavaria.* I unwrapped my wax-papered treasure and compared it to the pictures. The truffle, living its whole life cycle underground, was a different species altogether—*Rhizopogon*— actually a false truffle. Eyes blurring with fatigue, I read the descriptions of each fungus, and in print almost too fine to see, in a footnote at the bottom of each one, was "mycorrhizal fungus."

I flipped to the glossary. A *mycorrhizal fungus* formed a relationship— a life-or-death liaison—with a plant. Without entering into this partnership, neither the fungus nor the plant could survive. All three of my odd-duck mushrooms were the fruiting bodies of this group of fungi, which gathered water and nutrients from the soil in exchange for sugars made through photosynthesis from their plant partners.

A two-way exchange. A *mutualism*.

I read the words again, fighting the need for sleep. It was more efficient for the plant to invest in cultivating the fungi than growing more roots because the fungal walls were thin, lacked cellulose and lignin, and required far less energy to make. The mycorrhizal fungal threads grew between the cells of the plant roots, their spongy cell walls pressed against the thicker plant cell walls. The fungal cells grew in a web around each plant cell, like a hair net covering a chef's head. The plant passed photosynthetic sugars through its cell walls to the adjacent fungal cell. The fungus needed this sugary meal to grow its network of fungal threads through the soil to pick up water and nutrients. In return, the fungus delivered these soil resources back to the plant, through the layers of pressed-together fungal and plant cell walls, in a two-way market exchange for the photosynthetic sugars.

Mycorrhiza. How would I remember that word? *Myco* like fungus, and *rhiza* like root. *Mycorrhiza* was *fungus root. My. Core. Rise. Ah.*

Ah, yes. In a class I'd taken on soils, the professor had mentioned mycorrhizas so briefly, so in passing, that I hadn't taken any notes. He was teaching an agriculture class, not a forestry course. Scientists had recently figured out that mycorrhizal fungi helped food crops grow because the fungi could reach scarce minerals, nutrients, and water that the plants couldn't. Adding fertilizers full of minerals and nutrients, or providing irrigation, artificially took care of things, causing the fungi to disappear. When the plants didn't have reason to spend energy investing in fungi to meet their needs, they cut off the flow of resources. Foresters hadn't considered mycorrhizal fungi all that helpful to trees, at least not enough to teach about it, but a little thought had gone into inoculating nursery-grown seedlings with fungal spores to see if they helped the new shoots grow. But inconsistent results made it far easier to pour on fertilizer than cultivate healthy mycorrhizas. I chuckled at the humanness of this—we are always looking for the quick fix.

With a little effort, we could apply a more sustainable method by encouraging the development of the highly coevolved mycorrhizal relationships. Instead, foresters ignored the mycorrhizas, or—worse—killed them with fertilizer and irrigation in the seedling nursery, and focused only on those fungi that damaged or killed big trees, the *pathogens*. Those parasitic fungal species that infected roots and stems, damaged wood, and sometimes killed trees. The pathogenic fungi could, in short order, cost the industry a big whack of money. The professors in forestry school also taught us about *saprophytes,* the fungal species that decompose dead stuff, because they were obviously crucial to the cycling of nutrients. Without the saprophytes, the forest would choke from accumulated detritus, as our towns and cities would from garbage.

But compared with pathogenic and saprophytic fungi, mycorrhizal fungi were just not considered important. Yet they seemed to be the missing link between the life and death of the seedlings suffering in my plantations. Planting seedlings with naked roots in the soil was insufficient. The trees seemed to need helpful fungal symbionts too.

On my floor mattress, my back straight against the wall, I stared at my three prehistoric-looking mushrooms. They were *helpers* to plants, mycorrhizal fungi. That was what my mushroom book was telling me. I read a little further and found another startling passage. The mycorrhizal symbiosis was credited with the migration of ancient plants from the ocean to land about 450 to 700 million years ago. Colonization of plants with fungi enabled them to acquire sufficient nutrients from the barren, inhospitable rock to gain a toehold and survive on land. These authors were suggesting that cooperation was essential to evolution.

Then why did foresters place so much emphasis on competition?

I read the paragraph over and over. The bare roots of the yellow seedlings in the clear-cuts were trying to tell me why they were sick. The coral fungus with its cloud of spores, and the puffball mushroom with its fluttering hyphae, might have answers. So might the yellow gossamer webs on the root tips of the subalpine firs. I'd scanned the book the previous weekend to identify the pancake mushroom as *Suillus* but hadn't paid attention to whether it was a mycorrhizal fungus, a saprophyte, or a pathogen. I reread the *Suillus* description.

Suillus was a mycorrhizal fungus too. *A cooperator, a mediator, a helper!*

Maybe the fungus missing from the soil *was* the key to my dying seedlings. The industry had figured out how to grow seedlings in the nursery and plant them but totally missed that the collaborative relationships, the mycorrhizas, needed nurturing as well.

I wandered to the kitchen for a beer, grateful the boys had left a few cans of Canadian for the taking. The gas-powered fridge kept them cold, along with stacks of steaks and bacon. And the cheese and salami and iceberg lettuce loaded into the crisper. Loaves of white bread and tins of cookies for the next day's lunches lined the Arborite counter. The guys kept the space clean. I wished Kelly lived close enough for us to think this through together. He was probably already back in Williams Lake, set to start shoeing horses tomorrow morning, despite it being nearly impossible with his injury.

A moth beat its powdery wings around the flickering bulb hanging from my ceiling. A train whistled as it chugged along the banks of the Fraser River, the first of two journeying north along the gold-rush trail each night. I was glad I didn't work their shift. In bed, a threadbare sheet over my sticky knees, I sipped the beer and absentmindedly peeled the label. The puffball, coral, and pancake fungi could be helping the trees and one another. But how? I finished my beer and turned out the light, my brain churning while every muscle ached.

The dying seedlings had no mycorrhizal fungi, which meant they were not getting enough nutrients. The root tips of the healthy seedlings were covered in webs of colorful fungi, helping them acquire nutrients dissolved in water from the soil. This was mind-blowing. But I was missing part of the story and thought about today's clusters of trees. The old Douglas firs were clumped together in ravines in the deeply dry interior mountains. The soft-needled subalpine firs were huddled in clutches on mounds in the high-elevation mountains, as though they were escaping the frigid sodden spring soil. How did this clustering—whether growing down low or up high—help them survive? Maybe the fungi played a role in the grouping of trees in the most trying environments, bringing them together for a common purpose—to thrive.

The one thing I could be certain about was that I was onto something important that might fix the sickly plantations.

Somehow, the seedlings needed to become colonized by the mycorrhizal fungi to get resources from the soil. If I discovered more evidence, pushed in this direction, I'd have to convince the company to *change everything*. That didn't seem likely, given that I couldn't even persuade my boss, Ted, to plant a mix of species in the new clear-cuts at Boulder Creek. If cooperation, not competition, is the key to survival, how could I test this?

I pushed up the sash of the cracked window above my bed to let in the breezes rolling down from the precipitous mountain behind the bunkhouse. They blew in the scent of the trees and the sound of the creek and bathed my arms. Kelly's shoulder was hurting, and his hands were sore from gripping that rope, hanging on for dear life. What is it about pushing our limits that makes us stronger? How does suffering strengthen the relationships that hold us together? I loved the generous rhythm of the way the land and the forest and the rivers came together to refresh the winds at the close of each day. Helped settle us all down for the night. Air purified by the ancient forests hovered, and I let the downdraft cleanse me.

4

TREED

It was my twenty-second birthday, and my heart was set on celebrating in one of the wildest mountain forests of western North America. Kelly's shoulder was completely healed after only a year, and he was back on the rodeo circuit. My friend Jean was with me today, and we had our sights on the alpine up Stryen Creek, the first southern tributary of the seventy-five-kilometer Stein River, which flowed east into the enormous Fraser River at Lytton, British Columbia. We were only sixty kilometers south of my company town of Lillooet, which stood a thousand kilometers southwest from the Fraser River headwaters in the Rocky Mountains and more than three hundred kilometers northeast from its terminus at Vancouver on the coast. I felt drawn to this place, its mysterious energy. Jean and I had met in May, having both snagged a summer job with the British Columbia forest service, when I'd taken a break from my cut-and-run company and she was doing the same from another logging outfit on the Queen Charlotte Islands (Haida Gwaii). She'd noticed me in our university classes, but I was so quiet she'd assumed I was a francophone exchange student. We'd both been fortunate this summer to join the team of ecologists helping the province of British Columbia catalog the plants, mosses, lichens, mushrooms, soils, rocks, birds, and animals in the southern interior plateau, using the government's ecosystem classification system. Only a few months into the work, we'd already learned hundreds of species.

We were at the mouth of the Stein, where the canyon of white water joined Stryen Creek before flowing into the Fraser River. I was agi-

Jean, twenty-four years old, working in the bush near Lillooet, British Columbia, 1983. The hip chain around her waist was for measuring distance between plots for counting the number of regenerating trees. The trees at the edge of the landing are trembling aspen and those upslope are Douglas fir. This is the same truck that got mired in the mudhole when I was assessing the little yellow seedlings.

tated because of plans to log the Stein watershed over the next decade, and I'd already witnessed clear-cutting from one end of a valley to the other. I'd followed behind loggers, writing prescriptions to repopulate the clear-cuts, one blending into the next, with tiny seedlings, increasingly panicky because I was in love with forestry but furious at what was happening. It was in this state of confusion that I debated joining a protest the next weekend on Texas Creek, a northern tributary of the Stein. If discovered, I risked being fired.

Jean spread a topographic map across the hood of her Beetle. The main valley was narrow, rocky, and riverine, inlaid with braided trails worn from thousands of years of foot travel by the Nlaka'pamux people. "I've seen pictographs there," Jean said, pointing to a waterfall on the map. "They paint them with red ochre. Wolves and bears. Ravens and eagles. Young people coming of age go to the waterfall to sing and dance, and a guardian spirit in the shape of a bird or animal will visit their dreams. They gain endurance, strength, and immunity to danger and can change into other forms, like deer. There's a story that when a person becomes a deer, the tribe can kill and eat him, and if his bones are thrown into the water, he'll transform back into a person."

"No kidding." I looked at her in awe. "The deer are really people?"

"Yeah. The Coast Salish people think trees have personhood too. They teach that the forest is made of many nations living side by side in peace, each contributing to this earth."

"The trees are like us? And they're teachers?" I asked. *How did Jean know this?*

She nodded. "The Coast Salish say that the trees also teach about their symbiotic nature. That under the forest floor, there are fungi that keep the trees connected and strong."

I kept how startled I was to myself, but I couldn't imagine a more magical birthday present than to hear that my suspicions about fungi were already ingrained in those deeply connected to the natural world.

Jean packed the thin nylon rope for hoisting our food up a tree out of a bear's reach—we'd brought along wine, porridge, tuna-rice casserole, and a package of chocolate cake to bake over the campfire—and I had my plant guide. We tightened our hiking boots and hoisted our thirty-or-so-pound packs onto our backs. I cinched my shoulder straps—they were padded with duct tape and already hurting like hell—and tightened my hip belt. We needed to make it to the alpine before dark.

Not far from some ponderosas were switches of bluebunch wheatgrass—seeds alternately clasping each side of the main stalk, like hands climbing a rope. Feathery Queen Anne's lace grew knee-high in scattered bunches to cope with the aridity. Hearing the Indian story of the networks among trees got me wondering if the grasses, flowers, and shrubs along this trail might be mycorrhizal too. All but a few of the world's plant species—such as those grown on farms that are either naturally nonmycorrhizal or are irrigated and fertilized—require the helper fungi to soak up enough water and nutrients to survive. I pulled out some bunchgrasses with their pale blue-green sheaths, and a thick fan of rhizomes fell loose as I squinted at the root tips, hoping for the fat, colorful fungi I'd seen on the roots of healthy tree seedlings. But these looked naked; they were only fine, fibrous mops. When I checked a tuft of tall fescue grasses, the hairy awns of their seeds tickling my forearms, their roots were also bare. Same with the rhizomes of the spiky junegrass. Disappointed, I threw the grasses onto the trail.

We climbed to some widely spaced Douglas firs, their branches spread with the grandeur of oak trees. This part of the forest was moister. Pinegrass grew thickly under the fir crowns, with brighter, greener, more plentiful blades than the bluebunch wheatgrass we'd left behind with the ponderosas. I gripped a cluster of shoots, and the reddish stems of the pinegrass suddenly let loose. I fell on my pack— like an upended turtle. More skinny, scraggly, fibrous, naked root tips. They did not look mycorrhizal at all.

"What the heck are you trying to do, mow the lawn?" Jean said, grinning.

"I'm looking for mycorrhizas, but these roots all look bare," I said.

Jean tossed me a metal-edged hand lens the size of a monocle, and I squinted at the magnified roots. "They look kind of chubby," I said. "But not like the mycorrhizal root tips of the Douglas firs." I found the description for pinegrass in my plant book. A footnote read *arbuscular mycorrhizas* and said they could only be seen if stained with dye and checked under a microscope.

I flipped to the page showing Douglas fir. The footnote said *ectomycorrhizal.*

I stared at the grass roots in my hands, like a clump of hair pulled in a fight, and wished I could see *something* growing on the tips. I could have sworn they looked swollen.

"No wonder I'm confused," I moaned to Jean, scanning the pages. The arbuscular mycorrhizal fungi of grasses only grow *inside* the root cells. They're invisible. Not like the ectomycorrhizal fungi, which grow on the *outside* of the root cells of trees and shrubs, like tuques. The sun was high, and we needed to keep going or we'd be lost in the dark. But I couldn't believe what I was reading. "This is kind of gross. The arbuscular mycorrhizal fungus grows straight through the grass cell wall and permeates the insides where the cytoplasm and organelles are. It's as if they grow through skin and invade guts."

"Like ringworms?" asked Jean.

"Not really. A mycorrhizal fungus isn't a parasite; it's a helper," I said, explaining that inside the plant cell, the fungus grows in the shape of an oak tree. "Well, it forms a wiggly membrane shaped like a tree crown."

Putting her finger in the air as if she were dear Watson, Jean suggested this was why they were called *arbuscular* mycorrhiza. "A tree is an arbor," she said. "But why are the mycorrhizas on grasses different from those on trees?"

I shrugged. My book said that the arbor-shaped membrane has a huge surface area so the fungus can trade phosphorus and water with the plant for sugar. Good for helping plants in dry climates and where soils are low in phosphorus.

I threw the roots into the pinegrass swards, and we climbed past the stately Douglas-fir forest to where the trail flattened along a plateau. Except for the odd prickly spruce and leafy-green Sitka alder in the understory, the forest was completely taken over by skinny lodgepole pine, so named for its poker-straight stems and its excellence as a lodgepole for holding up a roof. Their boles were branchless and the high crowns small and tight, shying away from close neighbors.

I picked up some charred wood, surprised at how hard but light it was, as though petrified. Probably a remnant from a fire that would have opened the cones and spawned this thicket. Lodgepole-pine cones only open when the resin holding shut the scales starts to melt. These mountain forests burn every hundred years because of the cool but dry climate and frequent lightning strikes, combusting the whole stand and consuming the overstory. The scattered alders help replenish the nitrogen gassed out by the wildfire. They do this by supporting special symbiotic bacteria in their roots that convert the nitrogen gas back into forms that the plants and trees can use. In the absence of recurring fire, the light-loving pines would naturally die out in a hundred years, and shade-tolerant spruce would eventually dominate the canopy. The natural succession of things up here.

Fat huckleberries were thriving in bushes among the pinegrass, and I examined root tips here as well, only to find them also barren. Their mycorrhizal helpers formed yet another group: *ericoids*, fungi that form coils inside the plant cells, reminding me of the pin curls my mum used to set in my hair. Farther along, a ghostly, waxy white plant with translucent leaves and a hooded head was like a gleaming sword thrust among the shrubs. After a few minutes with my book, we identified it as a ghost pipe plant that parasitizes green plants because it has no chlorophyll itself. And it forms its own special kind of mycorrhiza—*monotropoid mycorrhizas*. We managed an exclamation between a laugh and a groan because it was yet another type.

How many were there? Monotropoid mycorrhizas are like the ecto-mycorrhizas, in that they form a fungal cap on the outside of the root tips. But they also grow inside the plant cells, like the arbuscular and ericoid varieties—making them perhaps a type in between. The ghost pipe plant's mycorrhizas also grow on the tree roots and steal their carbon.

Jean teased, "Aren't fungi what the French mainly eat? Even hal-lucinogenic mushrooms? You're hallucinating." She made a comment about the bottle of wine getting heavier, but she was beaming as broadly as I was.

With one thousand meters (three thousand feet) of elevation and ten kilometers (six miles) distance behind us, we reached the first rockslide. Scouler's willow and slide alder cascaded down the scree, making a good habitat for bears. Plenty of sun streamed over the towering arête. At the base of the slide was a miner's cabin, home to a gaggle of mice, rats, and squirrels. The single room was made from pine poles banged together with nails, and a little area had been cleared for a garden, probably to grow potatoes and carrots. Or maybe for burying dead people. Skin-crawlingly eerie, but we were starving. "Slap-slap cheese sandwiches," said Jean, handing them out. We'd per-fected the art of making a durable sandwich in seconds with cheese and pumpernickel bread. Just as I was thinking the place was too spooky, with ghost pipes creeping toward us from the forest's edge, Jean announced, "Probably some of those old gold miners died here."

She had a knack for saying things like that right as I was trying to swallow.

We set out again over dozens of switchbacks. At a zigzag, the mist of a waterfall showered us, and long-haired mosses draped the rocks. Skinny young lodgepole pines grew more sparsely and were slowly replaced by older subalpine fir and Engelmann spruce. By midafter-noon, in a hanging valley high up the mountain, the last switchback delivered us onto a flat where a creek tumbled down the bluff. We opened our arms at the top of the waterfall to feel the cool air rushing over us and the rock wall below. Jean pulled out her binoculars and said, "Look." The alpine was only a few hours away.

I scoured the landscape. Brilliant meadows swept toward snow-capped crags a few thousand meters above us. Fingers of subalpine firs, their crowns tapered by snow and strong winds, gradually dwindled into nothing among the alpine rocks. Closer to the creek grew thicker patches of subalpine fir and Engelmann spruce, with saplings regenerating in gaps created by crushing snow, lightning strikes, and gusts of wind.

"That's where I want to spend my birthday," I said, pointing to the ridges.

Along the roaring creek, crowded with thickets of leafy-green alders and supple willows, the trail was sketchy. It looked as if it hadn't been used in a long time. We tried to walk quickly, but the trail had other ideas. Muck coated our boots and trapped us in the dips. Logs crossed our path every ten or so meters and forced us to crawl over or wriggle under them, and devil's club stems scratched our arms. Rounding a bend, Jean halted at a bear scat the size of a turkey platter. "Grizzly," she said. "Blacks' aren't this big."

The scat glistened with huckleberries and grass. We kept shouting and winding through the alders and willows and found another scat. Even bigger and fresher.

Jean touched it. "Cold but soft," she muttered. "About a day old."

"I'm getting nervous," I said. Plus the creek was loud, and bears wouldn't see us come around a corner with all these bushes. Jean had already saved me earlier in the summer, after the tide trapped us at Tsusiat Falls on the West Coast Trail along Vancouver Island and we risked being swept into the ocean. I wasn't strong enough to climb the ten-meter-high cliff, so she'd tucked me under one arm and pulled me up with my pack on, probably 150 pounds all told, all the way to the top.

"Let's go a bit farther. I really want to have my birthday in the alpine," I said, but after the next bend, my chest tightened. Prints in the mud were as deep as my ankle and as long as my forearm. Claw marks, deep gouges, were a finger length from the end of the toe imprints.

"Grizzly for sure," Jean exclaimed. "These marks are huge. And look at the trees."

Fresh claw streaks were carved into the cottonwoods straight as arrows lining the creek. Five straight wounds ran parallel, each a meter long. Clear sap ran out of each fresh white scar like blood from a wound. A two-meter-tall cow parsnip, its toxic chemicals oozing from frayed leaves, lay uprooted. For the first time since I'd met Jean, I could see that she was scared.

"Come on!" I shouted. We could stay in the miner's cabin. No question, we had outstayed our welcome. I unhooked the bear horn from my belt as we raced for the switchbacks, heavy packs jostling. We didn't bother to adjust our shoulder straps for the downhill. By early evening, we reached the cabin—more ramshackle than I remembered, with gaps between the lodgepoles and ragged plastic covering the windows and doors. It was still safer than our tent.

We shed our fears by stirring up the cake mix with water and powdered milk in Jean's frying pan, covering the batter with foil, and roasting it over Jean's backpacking stove, laughing as it bubbled over the edge of the pan. We celebrated under a starry sky with cups of red wine and warm hunks of the chocolate cake, and we sang "Happy Birthday" like wolves howling at the moon. The Nlaka'pamux say when a man is transformed into a wolf, he finds courage and strength.

We talked late into the night beside our campfire. Ever since our West Coast Trail trip, Jean had struggled with depression. We talked about the sadness and dread that can break a life, feelings I knew all too well when my parents' marriage shattered and I'd been gripped by my first unshakable melancholy. Confusion that messed up my thinking. Jean said she sometimes felt like her mother, who was in an institution to deal with her own illnesses. I refilled our cups, the rich wine coursing through our veins and making the stars glow. We talked about the little things for coping, rituals we shared. Listing tiny tasks, such as "get out of bed" and "brush your teeth," slivers of accomplishment. Riding bikes up steep mountain roads until you're too exhausted to feel anything. Hiking ridgelines in sunshine so brilliant you crack a smile. My struggles were easy alongside hers. I just wanted Jean to be okay.

Eventually we doused the fire and went back to the pitch-darkness of the cabin. With the faint light of our headlamps, we spread our sleeping bags in the lodgepole bunk. I zipped my bag and burrowed down deep, as though it would protect me from more than the cold.

The next morning, Jean organized our breakfast while I went to an emerald pool to wash. I scanned the trees for signs of grizzly, but all was quiet. A collection of maidenhair ferns, with their delicate black stems, grew from a patch of humus at the base of a rock wall covered in a cascade of licorice ferns. I splashed my face. Lady ferns grew in recesses of humus, and tiny oak ferns covered the rises in the shadow of the trees. Each had, like Darwin's finches, found a niche.

Overwhelmed by a strong rotting smell, I glanced around. The trees and shrubs were motionless. The ferns were serene. It occurred to me that the odor was cache—rotting meat that a grizzly had dragged in overnight.

I hurried to the cabin and shouted, "Jean! Let's get out of here!"

We hastily slung on our packs as the pale sun was rising over the skyline peaks. On the trail beside the pool, we encountered the leg bone of a deer.

We raced down the trail, singing at the top of our lungs. Within minutes, we were hiking past the lodgepoles—nerve-wracking because the skinny trunks lacked branches, and even if we could somehow shinny up, the crinkly bark would cut our legs. Possible hiding places jumped out at me. Every curve in the path, every creek that might be crossed, every low-lying branch presented a potential escape route. After an eternity along the pine stretch, the trail descended back through the expanse of taller Douglas firs.

With their big branches and soft grassy understory, the firs felt friendly and secure. Dry Douglas-fir forests are not favorite habitat for grizzlies—they prefer the high-elevation forests and alpine meadows in August because they're cooler and the berries are ripening. I relaxed and broke into an even lope with Jean.

Down, down, down, feeling the weight of our packs. The duct tape I'd used to jerry-rig the right shoulder strap was fraying, so I fiddled with it, barely noticing the grasses and flowers waving at me. All of a sudden, Jean cried, "Grizzly!"

A few meters away were a mother and two cubs, staring straight at us. I reached for the air horn, but it had fallen off somewhere.

The bears were as stunned as we were. They were so close we could smell the carrion on their breath. We slowly backed up to the nearest trees. Jean dropped her pack and started climbing a Douglas fir, finding footholds on its gnarly limbs. I grasped the scaly trunk of a neighboring tree while the mama bear squealed at her cubs. Using my head as a battering ram, I ploughed through the thicket of branches. Jean was scrambling a good five meters higher than I was managing, and I was anxious to match her pace. The grizzly could easily tear me down while I was low. Blood poured from gashes and scrapes on my face and arms. My tree shook with fear. Jean's tree was letting her race up its massive trunk, speeding into the canopy. In my haste, I'd neglected to drop my backpack and had picked a much smaller tree! When I reached as high as I could go, it was swaying back and forth, and I was afraid I would drop onto Mama and her cubs, now wandering directly below my tree.

After glaring at me, she sent her cubs up two ponderosa pines. Safely out of the way while she dealt with us. The orange trunks had no branches, but the cubs were light and their claws were sharp. Mama snorted instructions as they scrambled and came to rest in crowns soaring well above where we clung. Mama turned toward us and reared up on her hind legs for a better look. Grizzlies are known for their poor eyesight. When she decided we were indisposed, she wore a path back and forth between the four trees. Perched high while she

Eating breakfast, twenty-two years old, at the miner's cabin at Stryen Creek, 1982

called the shots, I thanked my lucky stars. With my toes locked into a whorl, my hands bleeding, I leaned into my tree to rest. The warmth of the bark and sweet smell of the needles momentarily calmed me. Jean caught my eye and nodded toward the cubs. Their black eyes, framed in blond crew cuts, were peering at us. Jean couldn't help but grin at them.

Hours crept by. I shifted my feet to ease the pain in my back and resettled my pack, worried we'd be clinging here all night. Luckily I was too dehydrated from the hike to need to pee. The cubs, I swore, fell asleep while Mama sternly kept us all in detention.

I wished I could sleep too, but I couldn't stop trembling.

My mind drifted to Mum, because the vanilla scent wafting from the ponderosa bark reminded me of her kitchen, and I was desperate to ask her how to get out of this pickle.

Jean's resplendent tree wasn't shaking like mine; either Jean was more courageous than I was—of which I had little doubt—or the tree was stouter. A true elder. Leading, commanding, dignified. Its crown deeper and more imposing than those of its neighbors. Providing shade for the younger trees below. Shedding seed evolved over centuries. Stretching its prodigious limbs where songbirds roosted and nested. And where wolf lichens and mistletoes found crevices in which to root. Letting—needing—squirrels to run up and down its trunk in search of cones to store in middens for later meals. And to hang mushrooms in the crooks of branches to dry and eat. This tree alone was a scaffold for diversity, fueling the cycles of the forest.

My arms wrapped more tightly around the trunk. Mama settled under the ponderosas as her cubs slept. My tremble reduced to a quiver, my terror to mere fright. In the safety of my tree, I felt myself slowly grafting to its bark and melting into its heartwood, astonished at how calm I'd become in its branches. A woodpecker hammered at an ailing tree nearby, sending bark flying as it carved a new hole for its family. Next door, a snag hosted a larger cavity. It looked like a woodpecker hole too, but bigger and rougher because that tree had started to rot and the hole's edges had frayed. Woodpeckers in it wouldn't be safe from predators. Something moved inside it. The white face and yellow eyes of an owl peeked out. It turned its head and let out a hoot, maybe to the woodpecker, maybe curious about the commotion. The woodpecker and the owl seemed to know each other. Neighbors sharing nests and warning signals. The old trees bearing witness.

The smoldering glow of the sinking sun washed over the trees. My thoughts drifted toward the remnants of birthday cake in Jean's pack. Mama had wandered over from the ponderosas and was nosing around it.

She snorted a command. *Scratch scratch.* Her cubs scurried down, and they bounded through the bush with their mother, leaves chattering in their wake.

Then—silence. The branches sagged under my weight, and I imagined they were wishing I'd get off them.

"Think they're gone?" I called out to Jean as quietly as I could.

"I dunno, but I'm hungry. Time to go." She started down. I shouted my worry, but Jean—sensibly enough—pointed out that we couldn't stay in trees forever.

I shinnied down, reaching the base right after Jean's boots hit the ground. She looked at my scraped-raw arms but was even more impressed that her cuts were deeper. "We're lucky they didn't smell our blood," she said, inspecting her pack. No teeth marks. She unzipped one of the side pockets big as an elephant's ear—a source of pride because they doubled the size of her pack—and we downed the leftover cake. "Guess they don't like chocolate." Jean insisted she heard rocks falling up in the valley, and that meant we were safe.

Her tree was stolid and serene as it watched us leave. I glanced at mine, its leader nestled under the crown of Jean's. I wondered if Jean's tree was the parent of mine, since most seeds fall close by, almost all dropping to the ground within a hundred meters. A few heavy seeds are carried farther afield, across creeks and hollows, by squirrels, chipmunks, and birds. The odd one catches an updraft and flies across valleys on a wing. But most seeds drop on the outskirts of a tree's crown. Jean's old tree was likely the parent of my mine. It seemed protective of it, of all of us. I tipped my hat in thanks and whispered that I would be back to learn more from her.

We ran, banging pans and shouting at the grizzlies that we were leaving. But even in the throat of danger, I was enveloped in a new sense of peace, in a visceral, tactile, overwhelming wisdom of the elders, the firs and ponderosas. I felt the connectedness of the forest that the indigenous people already understood so deeply. I'd cried at the cutting of the old trees after Ray and I had laid out the clear-cuts in the Lillooet Mountains, and condemning five-hundred-year-old trees still haunted me with sensations of guilt. The efficiency of clear-cutting felt brutally detached from nature, a discounting of those whom we consider quieter, more holistic and spiritual.

But I was here with Jean in the forest for a reason. The trees had saved us, and I wondered if I could help my company find a new way to harvest them while protecting the plants and animals. And the mothers of the forest. Perhaps we could become leaders in the industry. Cutting was not going to stop as long as people needed lumber and paper, so new solutions had to be found. My grandfather had reaped harvests while leaving the forest vibrant and regenerative, the mothers intact. He was never wealthy, but he lived in rich peacefulness with the forest, taking only what he needed, leaving gaps so the trees could come back. I was fortunate he'd shown me this. How to protect the forest while it provided us with wood to build our homes, fibers to make our paper, and medicines to cure our ailments. I wanted to be a new breed of silviculturist who honored this responsibility.

I RETURNED to the logging company the next summer and stayed on until the end of September because I'd graduated from the university, but I got laid off when snow came to the mountains early, grinding fieldwork to a halt. I wanted to complete the planting prescriptions and seedling orders, but Ted promised to hire me back the following spring, which I hoped would lead to a permanent position.

A week later, I ran into him in front of the post office in Kamloops, a city one hundred kilometers to the north, where Mum was living. Ted looked like he wanted to hide when I said hello and asked how he was managing the office work I'd left unfinished. With a nervous laugh, he told me the company had hired Ray to finish writing the silviculture prescriptions through the winter. He stared past me and didn't offer any reason.

Had I done something wrong? It wasn't on account of the protest in the Stein Valley, because in the end I hadn't gone. I'd told myself I could better solve the problems from inside the industry. Nor was this due to my performance, because it was well known I'd learned more about the ecology and silviculture of the forests than any of the other students—including Ray. Didn't I fit in with the guys?

Ted called the next spring to offer me back the seasonal silviculture job as he'd promised, but I turned him down. I wanted to find another way to work in the wild. A way I hoped would give me more insight into the mysterious manners of mothers in forests.

I had no idea this would require that I first learn how to poison a tree.

5

KILLING SOIL

"Suzie, I'm scared," Mum cried. We'd been picking our way across a rockslide, steep walls above us that only goats could navigate, with boulders strewn like cars in a pileup. I looked behind to see her on an enormous one, sliding backward toward a wide gap.

I leapt across the rocks, grabbed the top of her backpack, and helped her crawl forward. We were in the Lizzie Lake alpine of the high divide between the Stein Valley to the east and Lillooet Lake to the west. Mum had no experience climbing rockfalls despite growing up in the Monashee Mountains, and I wanted to kick myself. My feelings were raw from being passed over for the winter silviculture job, and I wanted to ask her advice while showing her the landscape I'd come to love. But did I have to imperil her in the process? She could have broken her arm.

"Let's rest, Mum," I said. She'd been sweating so much there were wet streaks on the leather panels she'd excitedly sewn over the holes in her backpack, just for this trip. I'd given it to her when I'd bought a bigger climbing pack like Jean's. I pulled out the trail mix, and she picked through it for the chocolate. It was nice to comfort her for a moment.

"I've hiked the West Coast Trail, Suzie," she said, "but I've never backpacked across a field of bowling balls."

"Yeah, it's hard to balance on a round stone with twenty pounds on your back," I said, pretending to be on a tightrope to show how tippy I knew it was. "You have to shift your pack as you climb so it's a ballast.

It's a lot like skiing. Keep shifting the weight in tune with the angle of the boulder, as if you're shushing through moguls." Mum had become a skilled skier after her divorce, and every year since she'd bought us a family pass to the local ski hill. The first day on the rope tow, she fell at each turn. By season's end, she could snowplow down the chairlift run. By the second year, she was paralleling down the mogul fields on the mountain, determined to be as good as her teenaged kids. She'd make huge lunches with homemade bread and cookies, then ferry our friends and us to where we'd ski in a pack like pups and a mother wolf.

"If I can ski down the Chief, I can hike across a boulder field," she said, tossing peanuts to a hoary marmot. "I love those big ground-hogs," she said, delighted as he ate. Across the valley were jutting graphite peaks shaped by glaciers and sweeping avalanches. Beneath them scrolled bands of clear-cuts running through the subalpine-fir woodlands up high to the Douglas-fir forests down low. The shrubs in the cuts glowed reddish orange on this Canadian Thanksgiving Day weekend in early October.

"What are those cute flowers, Suzie?" She pointed at silvery seed heads atop skinny stems with parsley-like leaves.

"Tow-head babies," I answered, stroking a seed head with my palm. Handfuls were growing in a wad of humus that had collected between two boulders, glistening in the sun.

"Baby tow-heads!" she exclaimed. I loved her mixed-up version even more. "I see why you brought me here, Suzie. It's a special place."

"It's sketchier through there," I said, pointing to big gaps that the stone cairn trail markers were leading us through.

"No problem," she said. "It's not the first time I've been hiking in the Stein, you know," she said. Feisty like Grampa Bert, and stubborn and full of resolve like Grannie Winnie. Such a wonderful mix of the two that Robyn, Kelly, and I later combined their full names, Hubert and Winnifred, into the nickname "Bertifred."

"You've been near here before?" I was still of the age that I felt I knew far more than my parents. But Mum never ceased to amaze me, a traveler to Europe and Asia, a reader of Aristotle and Chomsky, Shakespeare and Dostoyevsky.

"I hiked with friends to the Asking Rock at the mouth of the Stein River where it joins Stryen Creek," she said, tying a kerchief around her neck because her thick brown hair was short and she took care to avoid burning. "It's a huge rock with cradles hollowed by the water, where the Nlaka'pamux women give birth." They baptized their babies in the creek, and the Asking Rock was where they asked permission to enter the Stein Valley. For safe travels.

How had Jean and I missed this during our summer hike? Unnerving, the eerie possibility that she and I had been treed by grizzlies and chased out of the valley because we were ignorant of the order of things.

By afternoon, we'd set up our tent on a rock ledge. I hung our food high in a subalpine fir—clearly the parent of the young ones growing around its base—to discourage any bears. Below us shone Lizzie Lake like a jewel nestled in green velvet, and above beckoned a glacier outflow pocked with small alpine lakes. We spent the afternoon climbing over the scoured rocks and dipping our toes in the pools.

"Look at the lichen on this rock, Mum." A red pie-shaped crust was bordered with whitish fungal filaments radiating outward. A symbiosis. "A fungus took a likin' to an alga," I said.

She pursed her lips at my joke and said, "Kinda looks like the dried puke I cleaned up in the boy's washroom last week." Mum was a teacher, a remedial consultant in an elementary school, where she worked with little kids struggling with reading, writing, and math.

I exclaimed over another patch, a deeper bed of lichen-crusted humus on the rock, with white mountain heathers sprouting from the center. The tiny flowers hung like fairy bells on top of short, snaking stems clad in leathery, scaly leaves. The heathers looked happy in their bed of lichen soil. The roots of the lichens—the *rhizines*—exuded enzymes to break down rock, while the lichens' bodies contributed organic material, and together they made humus for plants to root and grow. I tugged at one of the heathers, but it was solidly anchored in the lichen-spawned humus. Would I find roots with a net of fungus attached? Or a truffle? I hated to ruin this oasis by looking for mycorrhizas, so I checked my plant guide. The heathers formed a symbiosis with the coil-like ericoid fungi, the same type I'd discovered on the huckleberries at Stryen Creek with Jean. These lichen-fungi turned rocks into sand and released minerals, slowly making soil that other plants could grow in.

When I read the passage aloud to Mum, she nodded. "Makes sense. It just takes one plant to get going, then the others come along." She pointed to larger islands of green that had manufactured thicker layers of organic matter over other rocks. Pink mountain heathers and crowberries had roots in the crust. Some even housed sprigs of shrubs.

"Dwarf huckleberries," I said, indicating some short stems growing in the lichen humus, loaded with tiny blue berries. This species only grew in the alpine. Not like the huckleberries at Grannie Winnie's. Mum and I picked a few, wandering from patch to patch.

"Grannie Winnie would know how to grow a garden here if she had to," I said.

Mum laughed. Her mother could grow something from almost nothing. She just needed seeds, compost, and water. "It's like teaching kids to read," she said. "Give them the basics, and bit by bit, they will learn."

"Mum, I feel sick that they gave my job to Ray," I blurted. "What should I do?"

She stopped her berry picking and faced me. "Apply for another job, Suzie," she said, matter-of-factly. "Pick yourself up. Use what you learned from the company—from that guy Ted—and don't look back."

"I don't understand it; I didn't screw up." I didn't want to let go of what I felt was an injustice.

"Maybe they weren't ready to hire you. You'll find something even better."

She was right. Why was I so impatient? Mum wasn't. She could build up the sounds of the alphabet with her students for months. She'd cared for us day by day, in little ways that added up. Come to think of it, the lichens and mosses and algae and fungi were also steady as could be, gradually building up the soil, quietly in tandem. Things—and people—working together so that something noticeable could occur. Like the way Mum and I visited, made the time to be in step together, each moment connecting us more closely until we were made whole—our love rich, and varied, and deep-rooted between us.

Mum smiled serenely as she stretched out to rest. Born dirt-poor during the Depression, she'd watched her dad return from the war with post-traumatic stress syndrome. She married a man who was good but wrong for her, had three kids by the time she was twenty-six, got her teaching degree through correspondence and summer school, held down a full-time job while raising her family at a time when women were expected to stay at home, taught reading to kids who were poor, abused, or otherwise disadvantaged, suffered headaches that could kill a horse, divorced my father against everyone's will, then sent all three of us kids to the university almost single-handedly. She'd been through a knothole sideways, but to me, she might as well have been the first person to walk on the moon.

AS SOON AS I GOT HOME from our hike, I dusted off my résumé and applied for jobs with logging companies.

I landed two interviews. The first involved sitting across the vast desk from a manager at Weyerhaeuser who told me he couldn't wait to cut down all the old-growth forest so they could reconfigure the mill for small plantation trees. At the second, the guy at Tolko Industries told me they were trying to mechanize as much as possible. Neither offered me a job.

"There's a new silviculture researcher at the Forest Service, named Alan Vyse. You should try him," Jean said when I dragged myself home from Tolko and flopped onto the brown davenport we'd bought at a garage sale. She and I were sharing an apartment in Kamloops in south-central British Columbia, the blue-collar pulp-mill town where Mum was also living, only five minutes away. Jean had just gotten a year-long job at the Forest Service investigating regeneration problems in the dry Douglas-fir forests.

"Or I can collect pogey," I said, counting the weeks I'd worked and hoping they'd add up to the magic number for collecting unemployment insurance.

"Alan's tough but very smart. You'll make a good impression," Jean said softly.

WHEN I WALKED into Alan Vyse's office, he smiled and shook my hand. His hollow cheeks and high-tech sneakers told me he was a serious runner. He beckoned me to sit near his oak desk, a neat pile of journal articles on one side and a partially finished manuscript in front of him. A shelf was loaded with books on forests, trees, and birds next to hooks holding his cruiser vest, rain gear, and binoculars, work boots set underneath. A government office with beige walls and a view over a parking lot, but it was comfortable and felt as though weighty conversations had taken place here. I glanced at the egg yolk dripped down the front of my T-shirt. If he noticed, he didn't let on. Although he had the countenance of a consequential man, his eyes were full of kindness. He asked about my experiences in the bush, my interests, my family background, my long-term goals.

I told him about my summer jobs and ecosystem classification work for the Forest Service, squaring my shoulders. "That's experience in industry *and* government," I said, hoping he'd agree it was a well-rounded background for someone who was only twenty-three.

"Ever done any research?" he asked, muddy green eyes piercing me, as if the bare truth were right behind my head. He'd honed in on the gaping hole in my résumé.

"No, but I was a teaching assistant for a couple of courses during my undergraduate degree, and I was once a research assistant at the Forest Service," I said, my voice straining so much I had the added task of working not to wince.

"What do you know about regeneration?" He scribbled notes on a yellow pad. Foresters in green pants and taupe shirts strode by, one carrying a shovel and the other a piss-tank—a backpack water can with a hand-held pump—for fighting fires.

I told him about my yellow seedlings in the Lillooet Mountains and how I wanted to understand why the plantations were failing. That I wasn't planning to return to the logging company to finish my quest didn't come up. But I told him I'd figured out that tinkering with different planting prescriptions would never answer my question, because it was impossible to isolate my root problem when so many other things were changing at the same time. I told him I'd tried ordering seedlings with bigger roots, planting trees in duff, and planting them near other plants with mycorrhizal fungi, in hopes the fungi would contact my seedlings.

"You need to understand experimental design to sort this out," he explained. He pulled down a worn text on statistics from his bookshelf, and I noticed his master's degree in forest economics from the University of Toronto framed next to his undergraduate degree in forestry from the University of Aberdeen. Alan had an English accent, but I guessed that he also had Scottish blood.

"I took statistics at the university," I said. Glancing at the award on his desk for his years of excellent service—a gold plaque etched with a tree and his name—I felt totally naïve. He put me at ease when he told me that neither degree had prepared him for designing experiments, so he'd had to teach himself.

He didn't have any jobs available but assured me that there might be contracts in the spring to investigate "free-to-grow plantations," and he'd give me a call.

I had no clue what "free to grow" meant, and I left wondering if I'd hit the end of the line. I didn't yet know it was a new government policy to get rid of neighboring plants so that conifer seedlings were "free to grow" without competition from anything not-conifer—meaning any native plants, which were viewed as weeds to be eradicated. A policy that had grown from the influence of more intensive American practices that increasingly treated forests as tree farms. And here I was talking about seedlings needing to grow near huckleberries and alders and willows. *What an idiot I am,* I thought. Why did I mention the little yellow seedlings? He'll think my world is so small, that they were all I cared about. It was November, and spring was so far off that even if he thought I might be worthy, he'd forget me by then.

I applied for a lifeguarding job at the swimming pool. If everything failed, I was qualified for pogey, though Dad wouldn't be happy about me collecting from the government. I ended up landing a part-time office job editing a government report on forests, and I skied in the backcountry and regretted not making the time to visit Kelly. But he was busy shoeing horses and delivering calves.

Alan called in February. He had found a contract project for me to investigate about weeding effects in high-elevation clear-cuts. It wasn't getting exactly at the problem I was interested in, but it would build my research skills. He'd help me design the experiment and mentor me through the study, but I'd need to hire help with the bush work.

I couldn't believe it. I called Mum, who said she'd bake two chickens to celebrate. "Maybe you could hire Robyn," she said, banging pans as she started dinner at once. Robyn's substitute teaching assignments were spotty, and she needed a summer job.

A brilliant idea. I called Kelly on the phone to update him, and he hollered, "*Jeezus Christ,* Suzie," just as Uncle Wayne would. "Great news!" He told me that Williams Lake was colder than a polar bear's ass, but his farrier business was going well. Even better, he had met a new girl, Tiffany.

ROBYN AND I arrived in Blue River, the closest town to our experiment, set to take place in the high-elevation Engelmann-spruce and subalpine-fir forests of the Cariboo Mountains just west of the Rockies. The town had sprung up a hundred years earlier to support the fur trade and the building of the railroad and Yellowhead Highway, a settlement that displaced the Nlaka'pamux people who'd lived there for at least seven thousand years. They'd been relocated to a small reservation where the Blue River met the North Thompson River.

What was I doing? I was in charge of an experiment that required me to kill plants, creating yet another type of displacement. My task suddenly felt contrary to all my aims.

The three-hundred-year-old forest had been clear-cut a few years earlier, and without the overstory canopy blocking the sunlight, white-flowered rhododendrons and false azaleas, black huckleberries and gooseberries, and elderberries and raspberries had grown in abundance. The shrubs had spread their branches and produced a sea of leaves, flowers, and berries. So too the herbs had run riot—Sitka valerian, paintbrush, and lily of the valley. Sharp-needled spruce seeds had germinated among them, and nursery-grown spruce seedlings had been planted later to augment this natural stocking. But the planted seedlings were growing only a half centimeter per year, far less than necessary to meet future harvest expectations. Many had died, and the clear-cut had been declared "not satisfactorily restocked."

To fix this problem, the company foresters planned to spray herbicides to kill the overtopping shrubs, thereby "releasing" the remaining planted prickly spruce seedlings so they could have all the light, water, and nutrients to themselves. Monsanto had invented an herbicide in the early 1970s—glyphosate, or Roundup—that would poison

Robyn, twenty-nine years old, working at Mabel Lake, ca. 1987. Weyerhaeuser Company ("Weyco") was hauling timber cut from the rain forest near Kingfisher Creek along Simard Forest Road. Robyn was working with Jean at the time, assessing problems associated with seedling regeneration in the cut-blocks.

the native plants without affecting the conifer seedlings. Roundup had become so popular that many people used it casually on their lawns and gardens, Grannie Winnie a stubborn exception. The idea for forest plantations was that killing the leafy plants would free the seedlings from competition, and the companies could then meet their legal obligations for "free-to-grow" stocking. Free to grow like the dickens and get clear-cut again in a hundred years, far sooner than if left to grow naturally like the previous stand. Once free to grow, the plantation would be considered a well-managed forest.

Alan had helped me design the experiment to test how effectively different volumes of herbicide would kill the native plants so as to "release" the understory seedlings from competition. Presumably so they'd survive better and grow faster, and so that stocking and height-growth standards, and hence the free-to-grow policy regulations, were met. This was the job Robyn and I aimed to accomplish in this clear-cut, despite my misgivings. Alan wasn't enamored with the new free-to-grow policy either, but it was his job to test whether killing the shrubs improved the productivity of the plantations. He'd already told me he thought this policy was wrongheaded, but we needed rigorous, credible science that started with what the government believed before we could convince anyone of how to make changes.

That meant figuring out, step-by-step, how different doses of herbicide affected the seedlings and plant community. And comparing whether we should use clippers instead, or do nothing. To see if killing the non-cash-crop plants did indeed create a free-growing plantation that was healthier and more productive than if the natives were left to flourish.

With Alan's help, I devised four weeding treatments, testing three volumes of Roundup, one, three, and six liters per hectare, plus one manual-cutting application. We also added a control, where we'd leave the shrubs untouched. We needed to repeat these five treatments ten times each so we could be sure which worked best. We randomly assigned the replicated treatments, one to each of fifty circular plots. A statistician gave a stamp of approval to the design we drew on a map. A whole new world opened up for me. With Alan's guidance, I'd designed my first experiment! Though I loathed its purpose, which I was sure was the opposite of what we should be doing, I felt one step closer to having the skills to solve the puzzle of my little yellow seedlings.

Robyn and I set up pup tents at our campsite on the municipal grounds of Blue River, hers orange and mine blue, pitched on opposite sides of the firepit. We needed a built-in escape from each other, because the experiment would take several weeks and we were made of the same cloth—protective of our own turf. I placed my knockoff gas stove on a cut round of wood, and Robyn put her pots and pans on the picnic table to complete our living space. She offered to make a huckleberry pie using Grannie Winnie's recipe. Robyn loved to cook, something she'd learned as the eldest daughter of a working mother. The secret to Grannie's pies was picking the sweetest low-bush huckleberries in mid-August, when they were deep blue with a whitish tinge. Then baking them in a crust with lots of butter. After less than an hour of foraging along the trails that wound through the town, we had two pails full. Robyn made the pie on my tiny stove while I grilled hamburgers over the fire.

After dinner, we wandered around town. She'd had a job the previous winter cooking at the Blue River Hotel, the historic two-story wooden building with a dining room, beer parlor, and guest rooms upstairs, and when we passed it, she said, "Everyone there loved my pies." After meandering back to our campground, Robyn lost herself in a novel while I wandered in search of more berries. When I pulled up the roots of a pine seedling, I was delighted to find a bouquet of purple and pink ectomycorrhizal root tips.

Over the course of the week, we set up the experiment. Following the map Alan and I had sketched, Robyn and I used compasses and a nylon chain to locate the center points of the fifty circular plots. Each plot was about four meters in diameter, roughly the size of a tetherball ring. The centers were ten meters apart, so the size of our grid, when all was said and done, was one hundred meters by fifty meters, or a half hectare. Once we'd laid that out, we spent the next week identifying and measuring the abundance of the plants, mosses, lichens, and mushrooms within each plot so we could see how effective our treatments would be at killing them.

A few days later, we headed out at five a.m. to spray the treatments. Rounding the last corner, I slammed on the brakes at a rope barricade. Three protestors waved placards, protesting our arrival to spray herbicides. One agile man knew Robyn from her days working at the Blue River Hotel. A lively discussion ensued before they accepted that our hope was for the experiment to show that herbicides were not needed and would discourage their future use. They let us through.

The moment I'd been dreading had arrived. I'd bought the liters of glyphosate over the counter at the farmer's supply store in Kamloops, unnerved that anyone could just go in and buy it, but I was glad that I'd at least had to apply for a permit to spray it on government land. Robyn's fear was partly subdued by her frown. I measured the amount of pink liquid we needed for the one-liter-per-hectare treatment, poured it into the blue-and-yellow twenty-liter backpack herbicide sprayer, and added water to get the appropriate dilution. I coached Robyn in putting on a gas mask and rainsuit, same as me. I was still her little sister, but the order of our relationship—and who was in charge—was temporarily reversed. She'd been responsible for me her whole life, and now it was up to me to make sure she wasn't poisoned.

Robyn put on her mask and tightened the strap. She looked squarely at me through the goggles, as though to say I'd better know what the hell I was doing. Her long black hair was pulled back from her dark, angular face, revealing her thin Québécois nose. "This is heavy," she said, groaning as she hoisted the awkward square tank—it weighed about twenty-five pounds—onto her back and unraveled the hose leading to the wand.

I showed her what I'd practiced using water in Mum's yard, telling her to pump the lever while she sprayed.

The logs and bushes that had been easy to navigate while measuring the plants suddenly seemed like an obstacle course. Robyn's glasses fogged, and she let out a muffled cry under her gas mask, "I can't see, Suzie!" Like a seeing-eye dog, I steered her to the first plot.

She waved the black wand, shooting deadly mist over flowering rhododendrons and complaining that it didn't feel right. She hated killing these plants as much as I did. And wearing a plastic suit and gas mask while toting a backpack full of poison was making her grouchy.

I told her I'd apply the six-liter treatments to the next ten plots, trying to alleviate the pain of what I was making her do.

At the end of the day, we headed to the Blue River Legion for beer. The walls were covered in purple shag, and locals occupied the split-vinyl stools. A barmaid delivered our foamless beers, and when Robyn politely suggested they were a little flat, she said, "We don't serve no milkshakes here, honey."

Over the next three days, we applied all the herbicide treatments precisely. A-plus. A couple of days later, we came back with clippers and applied the manual treatment to those ten designated plots, leaving the remaining ten as untreated controls. Now we had to wait a month before measuring how effective the treatments were at killing the plants. I loved learning how to conduct an experiment in the forest but hated turning these plants into ghosts. For a forest management purpose that I already felt was mistaken.

When we returned, the rhododendrons, false azaleas, and huckleberries in the highest-dose treatment had shriveled and died. Not just the shrubs, but all of the plants, even the wild ginger and orchids. The lichens and mosses were brown, and the mushrooms were rotting. Some shrubs were trying to leaf out again, but the new leaves were yellow and stunted. Berries that had been plump on branches had fallen. Even the birds weren't eating them. Only the prickly spruce seedlings were alive, their needles still pale and stunted, some dripping with pink, and all no doubt in shock from the sudden flood of light. Most targeted plants were also dead in the intermediate treatment, but some were still green because they'd been hidden under the leaves of taller plants when the spray hit. At the lowest dose, most of the plants were still alive, but injured and suffering. The stems of the shrubs that had been cut were already sprouting back and overtopping the seedlings. The best treatment for making a free-to-grow plantation turned out to be the maximum dose of poison.

On the verge of tears, wanting to know how the glyphosate had killed the plants, Robyn said, "I know what we did. But what happened?" She always carried the brunt of our emotional pains, bearing the injustices, wanting to fix them.

I stared at my feet, because both of us crying would hurt too much. These plants were my allies, not my enemies. I raced over the reasons for doing this to justify them in my mind. I wanted to learn to do an experiment. I wanted to be a forest detective. This was for the greater good, for ultimately saving the seedlings. I would have proof that this was a stupid practice and be able to tell the government to investigate other avenues for helping seedlings grow. I looked at a thimbleberry plant trying to survive, its stems bare as they hunched over some pale seedlings newly revealed, but all it had managed to do was sprout a tiny basal pincushion of yellow leaves. The herbicide wasn't supposed to harm birds or animals, because the poison targeted the enzyme only the herbs and shrubs produced to develop protein.

But the mushrooms had shriveled and died.

Our favorite chanterelles—gone.

In my bones I knew the problem with the ailing seedlings was that they couldn't connect with the soil. That they needed the fungi to help them do that. And even then, the seedlings up here would grow slowly anyway, because of the snow nine months of the year. And yet I was explaining to Robyn that what we were trying to do was kill plants, including some shrubs that were the hosts of the fungi that I thought could help the seedlings. Companies were going mad with helicopters to blanket the province with glyphosate. Maybe our experiment would show that this plan wasn't all it was cracked up to be.

Robyn said, "Isn't it obvious looking at this mess that it's godawful wrong?" It hardly seemed possible that anyone could decide that free to grow was such a great thing.

At the campsite that night, we felt too sick to eat dinner. I huddled in my sleeping bag, and Robyn was silent in her tent. Hard to say if we were feeling ill from the herbicide or from regretting what we had done to the plants.

Alan shook his head at the result—that the highest dose of herbicide was best at killing the plants. As solace, he pointed out that this evidence still had nothing to do with detecting if the killing plan would help the seedlings. All it did was prove that a heavy dosage got rid of the so-called weeds. There was no time for remorse; we still had a lot of work ahead to unravel the complex relationships between seedlings and neighboring plants.

NOW THAT I KNEW how to establish a "weeding" experiment, I got a larger contract to test herbicide doses and manual-cutting treatments to kill leafy-green Sitka alder, lance-leaved Scouler's willow, white-barked paper birch, suckering aspens, and fast-growing cottonwoods. To obliterate purple-flowered fireweed, bunches of pinegrass, and white-topped Sitka valerian. Native plants, including trees, that might impede the growth of coveted planted seedlings—prickly spruce, skinny lodgepole pine, and soft-needled Douglas fir. These three conifer species, especially lodgepole pine, were now planted in nearly every clear-cut across the province because they were lucrative, durable, and fast growing. And the quicker the pesky native trees and plants were killed and free to grow achieved, the sooner the company's obligation to tend the plantations would be met.

The enshrinement of the free-to-grow policy amounted to all-out war on native plants and broadleaf trees. Robyn and I had become reluctant experts at the hacking, sawing, girdling, and poisoning of deciduous trees, shrubs, herbs, ferns, and any other unsuspecting creature in the new forests of the province. It didn't matter that the plants provided nests for birds and food for squirrels, hiding cover for deer and shelter for bear cubs, or that they added nutrients to the soil and prevented erosion—they simply had to go. Of no concern was the nitrogen added to the soil by the leafy-green alders, now clear-cut and burned to make way for seedlings. Or that the bunchy pinegrass provided shade for new Douglas-fir germinants, which otherwise ended up baking in the intense heat of wide-open clear-cuts. Or that the rhododendrons protected the smaller prickly-needled spruce seedlings from hard frosts that were much more severe out in the open than under a jigsaw canopy.

No, the thinking was clear and simple. *Get rid of the competition.* Once the light, water, and nutrients were freed up by obliterating the native plants, the lucrative conifers would suck them up and grow as fast as a redwood. A zero-sum game. Winners take all.

Here I was, a soldier in a war I didn't believe in. That familiar guilt at being part of the problem nagged at me as we began these new experiments. But I was in it for the ultimate prize: to learn how to be a scientist so I could unravel what was ailing the planted seedlings.

"I HAVE A SORE THROAT," Robyn said. We were heading back to our hotel after spraying alders at Belgo Creek near Kelowna, a couple hundred kilometers south of Kamloops. We'd been up since three a.m. to beat the heat. Not only was it too hot by noon to wear the plastic suits, the spray would evaporate from the leaves of the plants before it had a chance to kill them.

"So do I," I said.

"Think it's the spray?"

"I doubt it. We've been using it all summer. Maybe we have heat exhaustion."

The doctor at the clinic was gentle and could tell we were scared, and he took us into the examination room together. "Your throat is really red," he told Robyn, "but your glands aren't swollen. What have you been doing?"

When I told him we'd been spraying glyphosate, Robyn awarded me a stare as he inclined his head while asking, "Were you wearing a mask?"

When I said yes, he asked to see it. I fetched one from the truck, and he unscrewed the black plastic caps and whistled. "No filters," he said.

"What?" I said, peering with alarm at where the filters were supposed to go. We had been breathing in glyphosate spray all day. Robyn gripped the counter, and my legs started to buckle.

"You'll be fine, your throats are just burned from the chemical," he said. "Drink milkshakes, and you'll feel better in the morning." He patted Robyn reassuringly on the shoulder and smiled at me as we tumbled out the door, but I was as freaked out as Robyn. By the time we sucked down large chocolate shakes, our throats felt cooler. By morning, the soreness was gone.

It was the end of August, and this was our last experiment. Robyn would be leaving in a few days for a job in Nelson—a small town in southeastern British Columbia near where Mum grew up—as a substitute grade-one teacher, and she was missing her boyfriend, Bill. She didn't quit on me that day, but this was the last straw. Neither of us would ever forget the gravity of the mistake.

All but one of the treatments would end up failing to improve conifer growth and, no surprise, native plant diversity was lowered. In the case of birch, killing it improved the growth of some of the firs *but caused even more to die*—the opposite of expectations. When the birch roots had become stressed by the hacking and spraying, they had been unable to resist the *Armillaria* pathogenic fungus living naturally in the soil. The fungus infected the suffering birch roots and from them spread to the roots of the neighboring conifers. Where white-barked birch was left untouched in the control plots, and continued to grow intermingled with the conifers, however, the pathogen remained subdued in the soil. It was as though the birch were fostering an environment where the pathogen existed in homeostasis with the other soil organisms.

How much longer could I carry on this charade?

Then my luck changed.

A permanent job as a silviculture researcher with the Forest Service opened. I applied along with four young men. A panel of scientists was flown in from the provincial capital to ensure the hiring process was rigorous and fair, and I couldn't believe my good fortune when I landed the job. Alan would be my direct supervisor.

Now I was free to ask questions I thought were important. Or at least questions that I could try to convince the granting agency were important. I could conduct experiments to solve problems based on how I thought forests grew. Not just test policy-driven treatments that seemed to undermine the ecology of the forest, worsening problems. I could build on my experiences to conduct science that could help us help the forest recover from logging. Done were my days of testing herbicide treatments. Now I could really figure out what seedlings needed from fungi, soil, and other plants or trees.

I won a research grant to test whether conifer seedlings needed to connect with the mycorrhizal fungi in soil to survive. I added the twist of exploring whether native plants helped them make those connections, which I proposed to do by comparing seedlings that were planted in diverse communities to those planted alone, in bare earth. My ideas for this project, and my success at winning the grant, were owed in large part to what was happening in forestry south of the border. At the time, the United States Forest Service was transforming their practices, driven by public concerns over forest fragmentation and threats to species like the spotted owl, and scientists were recognizing that biodiversity, including conservation of fungi, trees, and wildlife, was important to forest productivity.

Could a single species thrive on its own?

If planted seedlings were mixed with other species, would that make for a healthier forest? Would planting the trees in clusters with other plants improve their growth, or should they be spaced far apart, in checkerboard grids?

These tests, too, might help me get at exactly why the old subalpine firs up high, and stately Douglas firs down low, grew in clumps. They could help me understand whether native plants growing next to conifers improved connections to the soil. Whether the conifers had more colorful fungi on their root tips when growing next to broadleaf trees and shrubs.

I picked paper birch as my test species, because I knew from childhood that it made rich humus that should be as helpful to conifers as it had been delicious in my dirt-eating days. I was also intrigued that it seemed to keep root pathogens at bay. But birch was only a weed to the timber companies. To everyone else, it was a gleaming provider of sturdy waterproof white bark, shady leaves, and refreshing sap.

The experiment should have been straightforward.

Holy cats, was I in for a surprise.

I planned to test how three lucrative tree species—larch, cedar, fir—fared in different mixtures of birch. I picked these as my other test species because they are the natives in the primary unlogged forests. I loved the cedar for its long, braided leaves, the Douglas fir for its silken bottlebrush laterals, and the larch for its starlike needles that turned golden before sprinkling to the forest floor in the fall. By now, the logging industry viewed birch as one of the most vicious of the competitors because it was thought to shade the coveted conifers, stunting their growth. But if birch saplings were helpful to the conifers, which mixtures would produce the healthiest forests? The three species of conifers differed in how much birch shade they could grow under, from very little for the star-needled larch, to lots for braided cedar, and somewhere in between for bottlebrush fir. This alone suggested that the best mixtures would vary with each species.

I settled on a design that paired paper birch first with Douglas firs in one patch, then birch with western red cedar in another, and western larch teamed with birch in yet another section of what was, at the time, a failed plantation in a clear-cut where not even lodgepole pines had managed to claim a home. I planned the same experiment in two other clear-cuts to see how the trees would respond in slightly different terrains.

In each species pairing, I planned for a wide variety of mixtures so I could compare the conifer species when they grew alone with when they grew with birch in different densities and proportions, and I could test my hunch that the mixtures would grow better in certain configurations, perhaps with low numbers of birch relative to larch, and higher numbers for cedar. I suspected that the paper birch enriched the soil with nutrients and provided a source of mycorrhizal fungi for the conifers. My earlier experiments also suggested that birch somehow protected the conifers against early death from *Armillaria* root disease.

This amounted to a total of fifty-one different mixtures, each on its own parcel. On three clear-cut sites.

After hundreds of days in plantations and in my weeding experiments observing how plants and seedlings grow together, I sensed that trees and plants could somehow perceive how close their neighbors were—and even *who* their neighbors were. Pine seedlings between sprawling, nitrogen-fixing alders could spread their branches farther than if they were hunkered under a thick cover of fireweed. Spruce germinants grew beautifully nestled right up to the wintergreens and plantains but kept a wide berth around the cow parsnips. Firs and cedars loved a moderate cover of birch but shrank when a dense cover of thimbleberry also grew overhead. Larch, on the other hand, needed a sparse neighborhood of paper birches for the best growth and the least mortality from root disease. I didn't know exactly how the plants perceived these conditions, but my experiences told me to plant the test mixtures with precision. Distances between trees had to be exact, and the clear-cuts had to be on flat ground for maximum accuracy. Given that British Columbia is a province of mountains, finding three flat sites would be no small feat.

To be as prepared as possible to look at the roots, to track if the conifers were connecting with the soil better when they were grown near paper birch than when they were alone, I ordered a dissecting microscope and a book on identifying the features of mycorrhizas and practiced with birch and fir roots collected on my way home. Jean would roll her eyes as I'd haul my samples into the storage-room-turned-office of our apartment, then tease me for burning the pot on the nights I'd promised to make dinner. My specialty was chili and hers was spaghetti, neither of us interested in cooking. I'd disappear into my cave-office until midnight, excising root tips, taking cross sections, and mounting them on slides. Before long, I was getting good at identifying Hartig nets, clamp connections, cystidia, and the many parts of the mycorrhizal root tip that helped distinguish one fungal species from another.

Some of the species of fungi on the soft-needled firs seemed the same as those on paper birch. If this was true, maybe the birch mycorrhizal fungi jumped onto fir root tips, cross-pollinating them. Maybe this co-inoculation or sharing of fungi or symbiosis helped newly planted Douglas-fir seedlings avoid having naked roots, perhaps letting them escape the same death sentence that befell my early yellow seedlings in the Lillooet Mountains. If fir somehow *needed* birch, birch wouldn't be hurting fir, as foresters assumed.

Quite the opposite.

After months of searching, I found three flat clear-cuts, all on government land—the sites of failed pine plantations, possibly because the soil biology was off-kilter. On one parcel, I ran afoul of a rancher who'd been grazing cows there illegally. He loudly protested my idea of converting the failed plantation into a testing ground, arguing that he had a right to a clear-cut he'd been homesteading for years. He was less than thrilled with my countering that as a research forester, I was entitled to the clear-cut and he was trespassing on public property.

Maudit tabernac! This was the last thing I needed.

Preparing to plant the experiment took another few months. It involved painting every one of the 81,600 planting spots on the ground. First, though, we had to deal with root-disease infections in all three clear-cuts. About twenty thousand old stumps from the original cuttings had to be hauled out of the soil because *Armillaria* root disease was infecting the dead roots and spreading as a parasite to surviving trees. About thirty thousand infected pines were dead, dying, or in terrible shape, so they had to be removed along with infected native plants. The forest floor became collateral damage from the excavations, ending up with huge piles of stumps, dead saplings, and diseased native plants bulldozed to the timber edge. But this left a clean slate.

I couldn't decide if the site looked like a farmer's field or a battlefield after the casualties got dragged off. My research grant didn't cover the installation of a cattle guard, so I painted a fake one across the road at the site entrance. I'd heard that cows don't cross lines on a road for fear they'll break their legs. It worked—for the first few months. The next summer, my crew and I spent a month in the hot sun, painstakingly planting the seedlings in their precise locations.

Within a few weeks, all of the seedlings were dead.

I was stunned. I had never seen such a complete plantation failure. I checked the rotting stems; there was no evidence of harmful sunscald or frost cankers. I dug up the roots and checked them under my home microscope. No obvious signs of pathological infection. But they reminded me of the embalmed spruce roots in Lillooet. No new root tips, just dark, unbranched sinker roots. Back at the site, lush swards of orchard grass had sprung up. I was puzzling over how this could have crowded in when the rancher drove up. "Your trees are dead!" he laughed, squinting at the wreckage.

"Yeah, I don't get it."

It turned out that he did get it. He very much got it. Furious at losing his grazing site, he'd seeded the clear-cut with dense grasses.

My crew and I (with under-the-breath mutterings, mostly mine) cleared out the grass and replanted the site. The plantation failed again—each of the mixtures. The white-barked paper birch died first, then the star-needled larch, then the soft-brushed fir, and finally the braided cedar. Following the order of their sensitivities to light and water shortages.

A third try the next year. Another failure.

A fourth replanting.

Again all the seedlings died. The site was a black hole where nothing would live. Nothing except luxuriant grass. The cows showed up to smirk at us, and I wanted to gather up all the cow shit and dump it on the rancher's truck. I guessed that the grasses had robbed the seedlings of water the first year, but I also had a troubling sense that the soil itself was suffering. I was quick to blame the rancher, but I secretly knew that my aggressive site preparation had displaced the forest floor and scraped the topsoil away. That couldn't have helped.

Douglas fir and western larch form symbioses only with ectomycorrhizal fungi, the ones that wrap the outside of the root tips, whereas the grasses formed relationships only with arbuscular mycorrhizal fungi that penetrate the cortical cells of their roots. The seedlings starved to death because the kind of mycorrhizal fungi they needed had been replaced by the kind only the damned grasses liked. It dawned on me that the rancher had helped me get at my deepest question: Is connection to *the right kind* of soil fungi crucial for the health of trees?

I replanted a fifth year, but this time I collected live soil from beneath old birch and fir trees in the adjacent forest. I placed a cup of it in each of one-third of the planting holes. I planned to compare these seedlings to another batch planted straight in another third of the razed ground without any transferred soil. For good measure, I placed old-growth soil that had been radiated in the lab to kill its fungi in the final third of the planting holes. This would help me figure out if the living fungi or the soil chemistry alone accounted for any seedling improvements with the soil transfers. After five tries, I felt on the cusp of a discovery.

I returned to the site the following year. The seedlings planted in the old-growth soil were thriving. As predicted, the seedlings without transferred soil, or with the dead, radiated transferred soil, were dead. They had met the usual morbid fate that had been plaguing them— and us—for years. I dug up samples of the seedlings and took them home to my microscope. As I expected, the dead seedlings had no new root tips. But when I looked at the seedlings grown in the old-growth soil, I jumped out of my chair.

Merde! The root tips were covered with a dazzling array of different fungi. Yellow, white, pink, purple, beige, black, gray, cream, you name it.

It *was* about the soil.

Jean had become an expert on Douglas-fir forests and the widespread poor seedling growth in dry, cold country, and I grabbed her to come look. She took off her glasses, peered into the microscope, and shouted, "Bingo!"

I was overjoyed. But I also knew I was only scratching the surface. Enormous clear-cuts had recently emerged on Simard Mountain, obliterating the old-growth forests. I had driven the new logging road along the shoreline where we used to moor Grampa's houseboat. Where the Jiggs outhouse used to be. And Grampa Henry's waterwheel, and his flume. Now one clear-cut morphed into the next. The cutting and monoculture planting and spraying had transformed my childhood forest. While elated with my revelation, I was heartbroken by the relentless harvesting, and it was my responsibility to stand up. To act against the government policies that I felt weakened the tree-soil links. The land. Our connection to the forest.

I also knew the religious fervor behind the policies and practices— a fervor backed by money.

On the day I left my experiment, I stopped to absorb the forest's wisdom. I walked up to an elder birch along the Eagle River where

I had collected the soil for transferring to the planting holes. Running my hands across the papery bark stretched across its wide, sturdy girth, I whispered the tree thanks for showing me some of its secrets. For saving my experiment.

Then I made it a promise.

A promise to learn how trees sense and signal other plants, insects, and fungi.

To get the word out.

The death of fungi in the soil, and the breakdown of the mycorrhizal symbiosis, held answers about why the little yellow spruce in my first plantations had been dying. I'd figured out that *accidently killing the mycorrhizal fungi also killed trees*. Turning to the native plants for their humus, and putting the fungi in the humus back into the plantation's soil, helped the trees.

In the distance, helicopters were spraying the valleys with chemicals to kill the aspens, alders, and birches in order to grow cash crops of spruces, pines, and firs. I hated this sound. I had to stop it.

I was especially puzzled by the war on alder, because *Frankia*—the symbiotic bacteria inside its roots—had the unique ability to convert atmospheric nitrogen into a form the small shrub could use to make leaves. When the alders shed their leaves in the fall and decayed, the nitrogen was released into the soil and became available for the pines to take up with their roots. The pines relied on this transformation of nitrogen because these forests burned every hundred years, sending much of the nitrogen back into the atmosphere.

But I would need much more proof about soil conditions and how trees might connect with, and signal to, other plants if I hoped to move the needle on forest practices. Alan had encouraged me to return to the university to get a graduate degree to keep improving my skills. I was twenty-six, starting my master's at Oregon State University in Corvallis in a few months, and I decided to conduct an experiment to test whether alder was a real pine killer, as believed by the policies, or whether alder improved the soil with nitrogen and gave pine a boost.

My bets were on the latter.

My hunch would prove to be far more prescient than I could have imagined. I knew my conviction to dig into free to grow could rankle the policymakers. I just didn't have any idea how much.

6

ALDER SWALES

By the time the prisoner transport truck arrived, I was having second thoughts.

Twenty prisoners in black-and-white stripes stumbled onto the logging road. Held in a corrections center north of Kamloops, they weren't murderers, more like thieves, but they were a rough bunch. The prison guard and a colleague from the Forest Service were quick to line them up. Robyn and I had a bird's-eye view from two hundred meters above in the clear-cut. She'd been my constant companion for more than a month, helping me establish my master's experiment in a decade-old clear-cut chock-full of Sitka alder swales.

The clear-cut was perfect for my current experiment, which aimed to examine how the shrubby alder influenced the survival and growth of lodgepole pine seedlings. Across the province, alders were being cut and sprayed into near oblivion so that pine plantations could be declared free to grow. This ambitious eradication program, costing millions, was being applied with zero evidence that it helped the pines grow, but it was a response—a dramatic one—to the fear that the alder shrubs suppressed and killed commercially valuable trees.

Alder grew in the understory of the native lodgepole pine forests, which had regenerated across the sweeping glaciated interior plateau following the fires ignited by settlers laying railroad and searching for gold in the late 1800s. A century later, these forests were clear-cut with feller bunchers—tractors with a mechanical arm wielding a saw—and the unlucky alders were either crushed by the wheels or cut along with the pines. With the overstory gone, light shined on the sheared-off alder stumps, which sprouted a multitude of new branches and leaves. Water and soil resources abounded. It was alder heaven. Clusters expanded handily from existing rootstocks, while below the leafy alder coronet, the pinegrass, fireweed, and thimbleberry were in full riot. To a forester driving by, the pine seedlings establishing within the sea of alder and grasses might seem to be drowning. I'd driven through many forests in the years leading up to my master's to see what these plantations looked like on the inside, getting out of my truck and wading into the alders crowding the road. Once through this wall of green, I'd usually discover pines growing beautifully. But seeing an ocean of alder from the roadside, even if many pines were poking through, was all foresters needed to justify a chemical assault or a literal hack job with saws or clippers.

But to what avail? No one knew whether this weeding was improving plantation growth. My experiment aimed to help fill that knowledge gap. I wanted to quantify the competitive effect of the alders and associated plants on the pines. Even more interesting to me was whether the native shrubs might actually collaborate with the pines to help them connect with the soil and create a healthy forest community.

To see how—and if—alder might interfere with pine, I had to reduce the abundance of the shoulder-height shrubs to different densities, including total elimination in some plots. Then I could compare pine growth with different numbers of alder neighbors to where pine grew alone, unencumbered by competition. Instead of merely thinning the alders, I decided to cut them all back and let them resprout in controlled amounts so that the ankle-high pines I was about to plant would face a realistic enemy. If they started the height race together, I could better gauge how they might compete on equal terms. If I'd been on the site with my experiment at the time of clear-cutting, I could have just thinned the newly sprouting alders as I planted my experimental pines. However, I'd come along after the alder stumps had already sprouted into full-sized shrubs. Nature is a dispassionate collaborator.

The prisoners were supposed to cut all the alders with machetes, leaving ankle-high stumps. Each alder shrub was a cluster of about thirty stems growing from a common rootstalk, a thicker version of the way a rosebush shoots up. To create the range of alder densities, my plan was to cut and to control which clusters could sprout new leaves, and which ones couldn't, by painting herbicide on the top of the stems of select clusters. In this way, we'd create five assorted densities—from one parcel with no alders (all of them painted with herbicide and killed) to 2,400 alder clumps per hectare (none of the alders painted, leaving all alive). We'd also create three more in-between levels (600, 1,200, and 1,600 clusters of sprouts per hectare).

Within the no-alder treatment, I also wanted to create a separate gradient of herbaceous cover—different amounts of pinegrass, fireweed, huckleberry, thimbleberry, and a dozen other minor species. I could then evaluate the competitive effect of this component on the pine seedlings separately from that of the alder. Alder was considered the main enemy, but the short plants were known to be fighters too. Oddly enough, only fireweed was a true herb, whereas pinegrass was obviously a grass, and huckleberry and thimbleberry were shrubs, but all were shorter than my knees, and so I lumped them together in what I called the "herbaceous layer." To evaluate the competitive effect of the herbaceous layer, I'd create three separate alder-free herbaceous treatments: 100 percent herb cover, where I'd leave the natural cover of herbs to grow freely; 50 percent herb cover, where I'd reduce the natural cover by half; and 0 percent herb cover, where I'd completely get rid of all the herbs. In each of these, I'd cut and paint the alder first, then spray herbicides to kill the assigned portions of herbs. In the total-annihilation treatment, I'd spray everything in sight—shrubs, herbs, grasses, and mosses—creating bare earth.

This extreme bare-earth treatment reminded me of the farm fields in the valley bottoms. A frightening battle plan, but I created it because American weed scientists in the 1980s, following in the path of the agricultural green revolution employing pesticides, fertilizers, and high-yield crop varieties, were finding that these conditions spawned the fastest-growing crops, and the British Columbia policy folks believed they could copy this to achieve the highest potential for pine growth. I would be remiss not to test their thinking that if they could get pines to grow like beans, they could harvest more forests in anticipation of faster growth. I needed to evaluate this against all other performance levels. We would repeat these seven treatments—

four with some alder retained plus three with no alder but different portions of the herb layer remaining—three times each. Each plot was twenty meters by twenty meters square, and all twenty-one plots were spread over about one hectare of the ten-hectare clear-cut.

In each of the seven treatments, I would plant pine seedlings so I could measure how intensely they'd compete—or collaborate—with the alder and herbaceous layers for light, water, and nutrients. I'd find out how much the alders helped the pine, maybe by adding nitrogen to the soil, and how much they competed for light, water, or other nutrients such as phosphorus, potassium, or sulfur. I would also learn whether the herbaceous plants were intense competitors or also protectors in some way. My aim was to log the amounts of resources that the pine, alder, and herbaceous plants would acquire. I'd also examine how fast the pines would grow and how well they'd survive across the seven levels of alder-and-herb abundance.

Robyn and I looked over the rolling montane pine forest as the gang marched up the trail. By now she was twenty-eight, two years beyond my twenty-six. She wanted me to signal that all would be fine, but I was skittish too. She was wearing a tank top, and I said, "I think you should—"

"Right," she said, pulling her lumberjack shirt over her low-cut top.

The prisoners grumbled as they walked toward us, forming a patter-song of curses. *This is fucked! I want a smoke!* They climbed over the barbed-wire fence with a lot of yelling. It was taut, five strands thick, and meticulous because Kelly, who'd built it on a week off from his farrier business, was aiming to keep cows out, not catch men in their crotches. *Goddammit, my pants ripped!* Muscles, chewing tobacco, long hair, hardened looks. "Look, chicks!" one exclaimed. "Hey, baby, wanna dance?" shouted another, gyrating his pelvis.

I explained to the guard how they should slash the shrubs to ground level. He listened, but things were potentially explosive. His only weapon was a baton. Robyn and I left them to their work and escaped to the far end of the experiment.

Our clippers and backpack sprayer were where we had left them. I'd decided she and I should establish the extreme 0 percent, bare-earth treatment ourselves, not involving the inmates. To make it as devoid of plant life as possible, we'd clipped every alder to a nub and hauled the stems to the side of the plots, leaving the grasses and herbs exposed. We coated the tops of the cut stems with 2,4-D and sprayed glyphosate over the grasses and herbs to kill the whole lot. In the 50 percent herb cover treatment, we'd sprayed only half of the herbs in a checkerboard arrangement. The sites looked barren. Neither of us felt good about killing the plants, but this time our greater purpose was more firmly in mind. If we found that these native plants weren't the killers the policymakers made them out to be, maybe the draconian practices across the province would be reconsidered.

We took off our plastic gloves and suits and rested at the edge of our last bare-earth plot. We'd been working since three a.m., our filters inserted properly. Robyn offered me a muffin made with huckleberries picked before the spraying. Even though we'd washed and were sitting outside the glistening plot, we ate with our hands in plastic bags.

"Look, voles," she exclaimed, pointing past water droplets of pink herbicide clinging to the leaves. They were scurrying, hauling clipped grasses to the mounds of alder branches we'd piled around the edges of the plot. "Bunny rabbits too!"

She hadn't yet grasped that the little creatures were eating poisoned blades. The scene flashed before my eyes: they'd share the deadly clippings with their babies in burrows, and they'd all die belowground.

"Go away," I yelled, running toward them. "Don't eat those!"

But we couldn't keep the voles, rabbits, and ground squirrels from browsing. We'd thrown them off-kilter by killing the alders. Robyn and I looked at each other helplessly—before we'd taken a single measurement, the disruption to the ecosystem was already evident.

And then we heard shouting. Robyn followed me to the plot of dense alder where we'd left the prisoners, a hundred meters away. The grunting was getting louder and faster. We snaked on our stomachs through a thicket of shrubs for a better look.

"Uhn, uhn, uhn," chorused the prisoners.

We witnessed the dissent. The men who were standing thrust their groins to the beat of the words. A guy with enough anger to frighten a ghost was leading the chant. Another with deep scars sat on a stump and chanted so forcefully that his neck veins ballooned. One skinny guy had a terrifyingly blank look. They had laid down their machetes in protest. The guard commanded the prisoners to stand, and Robyn and I held our breath. Anything could happen with only two unarmed watchmen, twenty prisoners, and the two of us.

The ringleader fell silent, and the guard and my forestry colleague marched them down the trail and into the van. They had been on the site only two hours.

When we inspected their machete work, I felt sick. I'd expected neat cuts to the bases of the alder shrubs, squared off so we could easily paint the herbicide to control the number of alders that would sprout back. Instead, the alders looked hacked to death, crowns slashed off, leaving sharpened stems as high as my thighs. Sap oozed from the torn bark and bled down the brown, speckled stems. The shoots looked like spears. A deer could impale its stomach on them.

A WEEK LATER, after Robyn and I completed the clipping we'd wanted the prisoners to do, the rest of my research assistants gathered. My family. Robyn, black hair in a ponytail, crouched next to the boxes of pine seedlings with her shovel, itching to put them in the ground. Kelly looked proficient in his jeans and cowboy boots, a carpenter's belt slung around his waist, ready to finish the gate and tighten the wires of the fence keeping out the cows—local ranchers had received a grazing permit. Jean, like a member of our family, carried the tools—calipers and measuring tape—for assessing the size and condition of the seedlings as they were being planted. Mum sat on a stump, notepad in hand, and smiled at her kids to convey how much she enjoyed seeing us like this. The edginess that Robyn and I had absorbed from the prisoners evaporated with Mum there. In a few weeks, Dad would arrive. Nicely staggered to avoid putting Mum and him in the same place at the same time.

Next to me was Don, the dark, curly-haired man I'd met in January at Oregon State University, a research assistant for a professor studying the effects of forest harvesting on long-term soil productivity. Don had taken me under his wing to show me the basics of being a grad student. How to use a spreadsheet, where to go running, where the best pubs were. "This is how you write statistical-analysis code," he had said, showing me something I'd long wondered about. I'd been counting the days until he showed up. He easily fit in with everyone, chatting about forestry. I was warm with excitement at having him close by. I was falling in love.

"Great job laying out these plots, Suzie," Don said, gazing at the posts we'd used to mark the corners of each of the twenty-one test sites. I was thrilled that he was already calling me by the nickname my family used. He reached over to touch the small of my back as he spoke.

To pull his attention back to the work, Robyn explained that each plot would be planted with seven rows of seven pine seedlings spaced two and a half meters apart, plus ten extras between rows that could be sacrificed for special destructive measurements. She could see he was competent—but she needed to make sure. Don didn't skip a beat, capping an explanation of how he'd proceed with one of his favorite Groucho Marx quotes: "Those are my principles, and if you don't like them . . . well, I have others." They were laughing as they took off up the hill to start planting the 1,239 seedlings among the various densities.

"Our turn," Jean said to Mum. They were to follow behind Robyn and Don. Mum jotted numbers on data sheets as Jean walked up the first row of freshly planted pine seedlings, pounding in wooden pickets with a metal marker beside each and then measuring heights with the ruler and diameters with the calipers. Beneath the candelabra of sprouting alders, the needles of the pine seedlings twirled in bunches like bouquets. These would eventually grow into the slender lodgepoles they were replacing, topped with tufted crowns like flames on a candle.

"You want a latched or walk-through gate, Suzie?" Kelly asked as we strolled to where he'd build the last section of the fence. I felt relieved to spend a few quiet minutes with him, studying the hummocks and dips where the posts would go. He'd added fencing to the long list of everything-cowboy he could do. He'd dug the holes by hand, his shoulders strong despite the dislocations from pursuing bull riding with the same passion he'd shown in his teens. He would end up creating a palisade around my experiment that would stand solidly for decades.

"I'm no expert, but a simple walk-through gate—you know, a

Y-shaped gap that a person but not a cow can wiggle through—would be good enough," I said.

"Yeah, that's easiest and cheapest," he said.

"It just needs to be big enough to haul our gear through, like the pressure bomb Dad and I will be using in a couple of weeks," I said, gesturing to show roughly how big the instrument was. About the size of Grannie Winnie's portable Singer sewing machine case.

"I can set the outer gate posts at an angle narrow enough that a cow can't squeeze through but wide enough for—a bomb?" he said, the scar stretching below his lip when he smiled. "I'd love to see Dad up here with the cows."

"It'll be fun. We'll be hauling the stuff in the middle of the night."

"Too bad he's not here today," Kelly said softly, still not over the breakup of our parents, even though it had happened thirteen years earlier.

"Way too tense with Mum here."

"I'll see him next weekend at the Williams Lake Stampede. I'm entered in the calf roping and bull riding."

"Right on," I said, thanking him for making such a skookum fence and asking him to say hi to Tiffany. I hadn't met her yet, but I'd heard she had wild red hair and could two-step like the dickens.

Kelly's face cracked into a radiant grin as wide as Mabel Lake as he said, "Thanks, I will." Proud of his work and touched that I appreciated it, he was still beaming as he picked up his shovel, waving at me to get on back to my trees.

Don helped Robyn and me plant more trees while Jean and Mum pounded in the marker stakes and collected the first seedling data. "Your landscape is so wild and empty compared to ours," he said, sweeping his arm across the view, making me proud of the land where I grew up. "I hope I can come to Canada to be with you always," he chattered. He was a talker. His words came in rushes, in rapids, and I loved it. Our days got packed with work that expanded from backbreaking into herculean. At night, I breathed in the earthy scent of his skin.

On his last day before returning to his research assistantship, Don and I collected the soil samples with a foot-long T-shaped soil corer. He showed me how to push the tip of the corer into the forest floor before pulling the handle to extract a long tube of mineral soil for each sample bag. Where we'd killed all of the plants, the soil was like butter. But extractions were damned hard where the plants grew sturdily, their maze of live carbon-rich roots resisting the corer and requiring us to stand on the handle to break the ground. By noon I was so sore he rubbed my back.

When he was set to leave, I cried. He reassured me that he'd be back in September to help with the final set of seedling measurements. We promised to go canoeing in Wells Gray Park.

Robyn and I returned a few weeks later to collect measurements of how much light, water, and nutrients were present in each treatment. Since the select alders were sprouting back and the remaining herbs were filling out their leaves, how much light did they preempt from the pine seedlings? How many nutrients did they absorb, and how much was left for the seedlings? How much water remained in the soil for the pine roots after the other plants met their needs?

To measure the water in the soil, we used a neutron probe. It was as deadly as its name implied, a yellow metal box that looked like a dynamite detonator—with a radioactive source of neutrons for measuring how tightly water adhered to soil pores. The scarcer the water, the greater its adhesion to the soil particles and the more difficult it was for the pines to take up—something the neutron probe could tell us. Alders, pines, and herbs all required water to carry out photosynthesis, but the alders needed the most in order to make enough energy to transform (fix) atmospheric nitrogen to ammonium, which the alders could then use. To accomplish this energy-demanding process, I expected them to suck up the most soil water. That was my hunch. The grasses and herbs, with their mat of fibrous roots, were also likely to be very thirsty.

Robyn and I carried the yellow box to an aluminum cylinder we'd drilled one meter into the ground and gingerly placed it on top. We'd inserted these tubes in each of the plots to measure the water levels in the soil. The more water alder took up, the higher its photosynthetic rate and the more energy it could invest in the nitrogen-fixing process. But at the same time, the less water it would be leaving for the pine seedlings. A trade-off.

Inside the box was a coiled cable ending in a housing tube containing a radioactive pellet that emitted neutrons. A plunger released the cable so that the housing tube descended into the cylinder, causing the pellet to shoot high-speed neutrons out to collide with water molecules in the soil. An electronic detector would register how many slowed-down neutrons boomeranged back as a measure of soil water content. With the push of a button, the cable would retract into the box as quickly as an electric cord into a vacuum cleaner.

"I don't have a clue how this works, but I want to have kids one day," Robyn said.

I plunged down the handle to release the cable into the tube. I hated the neutron probe. It was old and heavy, and the cable was sticky. I disliked the weird looks from drivers when the nuclear warning was on my tailgate. But more than anything, I was afraid of the radioactivity.

Measuring the water in the twenty-one access tubes took all day. We'd repeat this measurement several times over the rest of the summer to see how extreme the drought got in every plot, especially in the dense alder ones. It was finicky work because the instrument was awkward to carry, the cable didn't always descend properly, and sometimes the tubes were partly filled with water if the plastic Tim Hortons coffee cups we'd placed on top got tipped off by a squirrel.

At the last aluminum cylinder, relieved that this stressful day was almost done, I looked at the ground and gasped. The housing tube with the neutron source was dragging naked alongside our feet. The locking mechanism had failed, probably at the last cylinder, and the cable had not retracted. We were being zapped by the radioactivity.

"Suzie!" Robyn shouted.

"Shit!" I shouted back. I pressed the button on the yellow box, and the cable and the housing tube withdrew into the chamber.

How dangerous was this? As a condition of using the equipment, Atomic Energy of Canada had us pin a film dosimeter badge to our shirt pockets to measure exposure of our vital organs to the radiation. Feet are of less concern, since they have such a small mass and no organs and therefore would suffer little soft-tissue damage.

"I think we'll be fine," I said, explaining about the badges.

Robyn missed Bill, whom she was to marry in Mum's living room at Thanksgiving, and this blunder was the coup de grâce for her. I promised to send in the badges right away, anxious myself. Radiation causes cancer, after all. Jean distracted us at dinner by talking about the cows that had become bloated and were farting and burping after eating fertilizer accidently left in a plantation she'd been assessing. I wrote to Don, words pouring over one another about my fears.

When word came from Atomic Canada, I stared motionlessly at the results. Our exposures were well below the threshold considered a problem. Another narrow escape.

Robyn and I returned every two weeks from early June to late September to repeat the soil-water measurements with the probe. With the biweekly digital readouts, I analyzed the trends in soil water over the growing season. A distinct pattern emerged. In the spring, the recently melted snow had left the soil pores full of water. It didn't matter a hoot whether alder sprouted or not: no amount of alder could diminish the dampness left after two meters of winter snowpack had melted. But by early August, the soil pores had dried out where the alder grew back thickly. The flourishing alder leaves had been so eager to transpire gallons of water through their open stomata that they'd used up most of the free water. Where we'd completely eliminated the alder, however, the rootless soil pores remained full of water throughout the summer. Ugh, maybe the weed zealots were right. The alder did seem to be leaving very little for use by the pine seedlings in midsummer. The questions where the rubber hit the road: Did the alder-free pines grow fast as all get-out, as the policymakers expected, unlike the ones among the alders, and were they accessing and *using* the extra water when it was seasonally available?

To answer this, I needed to measure how much water was ending up in the pine seedlings in the middle of the summer. I enlisted my dad to help me.

WE LEFT TOWN AT MIDNIGHT on August seventh when, according to the readings Robyn and I had taken with the neutron probe, the alder-covered soil was the driest. It was a two-hour drive to my site. Dad's tall, thin frame, along with the huge lunch his new wife, Marlene, had packed, squeezed into the cab of my truck. He poured us coffee from his thermos as we trundled down the highway. As we bounced onto the logging road, the bushes more shadowy farther into the woods, Dad's stare deepened. He had never let his childhood fear of the dark dampen our family adventures, including weeks in the houseboat on isolated beaches at Mabel Lake, where anything could happen. I reassured him that I had bright lights that would make things easy.

Waiting for us at the timber edge was a pontoon-sized high-pressure cylinder of nitrogen gas. I'd explained we had to use this gas in the middle of the night to see whether the seedlings were recuperating from any daytime drought stress. I was guessing, based on the great amount of water in the soil measured by the neutron probe, that the pines in the bare-earth treatment would recover more fully in the night than those growing among the water-sucking alders. After the midnight measurement, we'd reassess the seedlings at noon to see how stressed they got in the heat of the day. If they were water-stressed in the day *and* the night, then I'd know which seedlings were in big trouble and could even die before the summer was out. It could explain why my seedlings in bare earth were starting to grow faster than those among the alders.

Dad fiddled with the rubber headband of his fist-sized headlamp. I turned the switch on high, and he immediately smiled at the flood of light. I did the same with mine and sparked up the two flashlights, then turned off the truck lights. We looked at each other. The headlamps and flashlights were no match for the deep dark. "Stick to me like a glue pot, Dad," I said. He nodded.

We had to transfer some gas from the big tank to a thermos-sized cylinder, because the big tank was too heavy to haul up the hill to the experiment, and I showed him how the regulator reduced the pressure of gas flowing from the big tank into the tubing connecting it to the little cylinder. Without the regulator, we'd blow the tubing and small tank, and maybe ourselves, to smithereens. I hid my nervousness over the thought of making a mistake.

We transferred the gas and set off through the inky bush, sticking so close together that we kept banging elbows. Dad carried the bear horn and small cylinder of nitrogen gas, and I lugged the twenty-pound pressure bomb. It would measure the amount of water pressure in the seedlings' *xylem*—the central vascular tissue in the stem that transports water.

When I shone my light on the gate, I said, "Kelly made this fence."

"He did?" Dad pulled the top wire to see how tight it was, then ran his index finger along it as if it were a piece of fine furniture. "It's perfect." Dad had worked hard to provide Kelly with everything, maybe to make up for his own childhood poverty. He gave him the best hockey gear and attended the games, and he signed him up for power skating and supported him on the all-star teams, wanting him to enjoy the ice, as many boys in Canada did.

We stumbled around, setting up a measuring station between a maximum-alder treatment and a bare-earth treatment. I placed a piece of plywood flat on an old stump and arranged the equipment on it. Dad stuck to me like Velcro. The heavy metal suitcase that housed the pressure chamber looked as if it might have held a bomb during the Cold War. Opening it revealed a chamber, a dial, and knobs, an apparatus that seemed equipped to detect a lie. Or electrocute a spy. "Pop had something like that," Dad said, and he whistled. Grampa's shop in the basement of his old house had been full of strange contraptions, mostly ones he'd built himself for logging.

"Go into that patch full of alders and cut a lateral branch from the marked pine seedling," I said, shining my light toward a pine festooned in pink tape. "Take the lateral branch, not the leader shoot, because without a leader, the pines won't know how to grow toward the sky."

Dad looked at me as if I'd asked him to jump off a cliff before whispering, "Got it."

He'd been pretty calm so far, but now I worried that his lifelong fear of danger in the bush—never mind in the pitch black—was so heightened that he might panic. "I'm right here, Dad," I said, turning on the Walkman I'd borrowed from Jean. Dire Straits' "Walk of Life" sang through the night and he vanished but for his bobbing headlight, while I kept calling out that I was right here. He returned moments later proudly holding a juicy leader shoot.

I took it anyway and stripped away the needles and phloem, leaving only the central xylem, an inch long. The xylem transports water from roots to shoots in response to the water deficit created by transpiration—the water vapor emitted from the stomata of the needles into the air during photosynthesis. During the day, water pressure in the xylem should be low as the roots struggle to pull water from the drying soil to meet the vapor deficit created by transpiration. The xylem pressure reading at night should be higher because the stomata are shut and the taproots are still accessing the ground water, leaving the xylem saturated and not under any water stress. If there is a strong midday drought, however, the seedlings might not completely recover in the night, and the cells in their xylem might still be dry at midnight.

I pressed the central xylem—all that now remained of the peeled stem—through the tiny hole I'd poked through the middle of a quarter-sized rubber stopper. The rest, shoots and needles, hung upside down from the bottom of the stopper. I inserted the rubber stopper into a quarter-sized hole drilled into the middle of the heavy screw-top lid of the pressure bomb, then stuffed the fluffy shoot inside the gas chamber, which was the size of a mason jar, and screwed the lid down tight. The tree looked like a bonsai hanging upside down inside a wide-mouthed wine carafe. I shined my light on the top of the screwed-down lid and was happy to see the piece of peeled xylem wood sticking straight into the air like a toothpick.

Dad stared with astonishment while I screwed the tube attached to the small cylinder of nitrogen gas tightly into the pressure chamber and twisted the knob to listen for the gas. A bubble of water would emerge from the twig's cut end when the pressure I applied equaled the resistance of the water held in the xylem. The more stress the seedling was under, the more tightly the water was held in the xylem and the more I'd need to turn the knob.

Dad's job was to yell, "Now!" when he saw the bubble.

Catching the excitement, he shouted, "Now!" so loudly I jumped. I snapped off the gas and whistled at the reading—five bars. The seedling was thirsty, not recovering fully at night. Dad assured me he'd collected the sample in the middle of the alder clump.

The alder was sucking up most of the water and leaving the seedlings dry. I explained that the alder probably needed tons of water to fuel the transformation of nitrogen to ammonium. The soil data also told me the alder was releasing a lot of nitrogen back to the soil when its leaves senesced and decomposed in the fall. The pine roots could then snatch up the released nitrogen. "This seedling should have lots of nitrogen in its needles," I said, "even though it's thirsty."

"Can we check that out?" Dad asked.

Agreeing to send these needles to the lab to get their nitrogen concentration, I opened the chamber and gave him the sample seedling. Within seconds, Dad had the needles in a plastic bag. I was starting to think he could make a good technician.

"Where's the next seedling to measure?" he said, eager for a return foray into the dark. This time he found a proper lateral branch, from a patch where all of the alder had been eliminated. The water-stress signal was zero. Its xylem was full of water. The seedlings where the alder had been removed were recovering at night because more water was in the soil. I tried not to be disappointed that so far the seedlings were proving the policymakers right: alder in midsummer was indeed claiming the water the pines needed. But my bigger inquiry aimed to address precisely the dangers of simple conclusions and shortsightedness. What would the long view—and the complexity added by the vital need for nitrogen—prove?

ROBYN AND I returned to measure soil water with the neutron probe three more times. Each time, Dad and I again followed at midnight to see how well the seedlings were responding to the water fluctuations in the soil.

I was surprised at what we discovered.

By late August, the neutron probe showed the soil under dense alder had filled with water again. *There was now—already—as much water in the dense-alder treatments as in the bare-earth patches.* Not only were the soil pores refilling with late summer rains and dewdrops, they were being inundated with groundwater at night, when the alder taproots pulled water from deep in the soil and exuded it through lateral roots into the dry surface soil in a process called *hydraulic redistribution.* Water rerouting.

And something else was happening in that denuded, pine-seedling-only soil. When the raindrops hit it, the water ran off the surface, carrying tiny particles of soil with it. Silt, clay, and humus grains got transported away in rivulets because there were no living leaves or roots to stop them. As the alder-dense plots began gaining water from late August and over the next few months, the bare earth started losing it.

Dad and I used the pressure bomb to test whether the seedlings were sensing the changes in soil water content. *As water refilled the soil, the stress the pines had experienced among the alders completely disappeared.* Except for that brief period in early August, the pines in the leafy-green alder patches were experiencing no more water stress than the ones in bare soil. It turned out that getting rid of alder so pines could be free to grow provided only a fleeting advantage in water uptake. All this killing was starting to look a lot like overkill. Not only that, its side effect was a loss of soil.

Next, I checked the light levels. The seedlings among the sprouting alder received almost as much sunlight as those in the bare-earth treatment, so improvements in light could not explain the faster growth rate of the alder-free pine. There was another major consideration: Don's soil samples showed that killing alder *halted* new nitrogen additions to the soil because the nitrogen-fixing bacterium *Frankia* had been eliminated when the alder roots died. Nitrogen is essential for the building of proteins, enzymes, and DNA, the stuff of leaves and photosynthesis and evolution. Without it, plants can't grow. It is also one of the most crucial nutrients in temperate forests because it frequently goes up in smoke in wildfires. Deficits in nitrogen, along with cold temperature, are known to limit tree growth in northern forests.

But while the addition of new nitrogen—or, more precisely, atmospheric nitrogen that had been transformed to ammonium—stopped with the loss of the alder and its partner *Frankia,* there was a short-term pulse of other nutrients (phosphorus, sulfur, calcium) to the soil as the dead roots and stems decomposed. As this detritus decayed, the alder proteins and DNA were further mineralized, or broken down, into the inorganic nitrogen compounds of ammonium and nitrate. Through these processes, the nitrogen was being recycled and released as inorganic nitrogen. The inorganic compounds, dissolved in soil water, were then readily available for the pine seedlings to take up, briefly boosting their growth. After about a year, however, after the dead alder had long since decomposed and the mineralized nitrogen had been consumed by seedlings or plants or microbes, or leached through the groundwater, the total amount of nitrogen in the bare-earth treatment plummeted compared to where alder grew freely. The short-term pulse of nitrogen as ammonium and nitrates released during decomposition was quickly used up, and there was no alder left to replace or augment that nitrogen. It went missing in action.

By fall of the first year, the short-term increases in water and nutrients—released during decomposition—resulted in enhanced growth of the planted pine seedlings compared to where the alders resprouted. This is what policymakers saw. But would the seedlings always be fine, or would the looming nitrogen shortage catch up to them? Reading the data felt eerily as if we were reading the seedlings' horoscopes.

"DO WE HAVE TO WAIT until this is a full-grown forest to get our answer?" Robyn asked.

I wasn't sure. I thought about the papers I'd read. It was clear that pines get nitrogen from the soil after it's been enriched by nitrogen-fixing plants like alder. The pine roots—or the mycorrhizal fungi colonizing their roots—then take up the nitrogen from the soil.

What I couldn't figure was why pines—the very foundation of a lodgepole forest—would be waiting for leftovers. Shouldn't they have figured out a better way to survive?

Perhaps those pine seedlings in the bare earth were able to acquire enough nitrogen after the dead roots of the alder and grass had decomposed. Or there might be a more direct source.

For now I'd plumbed the limits of my knowledge.

By October, I had the foliar nitrogen data in my hands. Pine seedlings growing among alder were rich in nitrogen, and those without it were depleted. Even though the pines among the dead alder roots in the bare-earth treatment were taking up more phosphorus and calcium released from decomposition, they were more depleted in nitrogen in particular because of the absence of new nitrogen additions to the soil. And although I lost a few of the seedlings growing among dense alders to midseason water stress, the rest were fine, chock-full of nitrogen and water and growing as rapidly as the ones in bare earth. This suggested to me that for most of the time, other than the most stressful weeks in August, alder was replenishing both water *and nitrogen* in the soil. How this forest functioned was turning out to be much more complex than the blunt free-to-grow policy presumed.

Policymakers, I thought, saw only the data of depletion. The short-term, the first roadside glance. Alder preempting resources that otherwise would be available to pine seedlings.

But once I stood back and took in the longer sweeps of time and season and scene, I could see that this was clearly not the whole story. It seemed the data were revealing a story of bounty.

Where I'd eliminated all of the alder, I also lost far more pine seedlings to voles and rabbits, which had made a beeline for the needles. The critters Robyn and I had worried about had reproduced like crazy in the clipped alder piles. The seedlings, the only green plants left in the bare-earth plots, were like magnets, and the rodents merrily clipped off the luscious shoots in the first season. All that was left of most of the planted seedlings were brown nubs. Rabbits left as clean a cut as Robyn and I had with our clippers. Still other seedlings succumbed to frost damage, leaving short yellow needles and eventually nothing but pale, dead stems. A number were sunburned, with scars at the bases where there was no shade—protection normally provided by leafy plants around them. By the end of the summer, more than half of the seedlings stripped of their alder neighbors were dead. Bare earth was looking as inhospitable as the moon.

Almost all the pines among the alders, on the other hand, were alive. They were growing at a slightly slower pace compared to the few remaining in the bare-earth treatment, but their needles were healthy and deep green. When I added up the volume of wood of the seedlings—all fifty-nine we'd planted in the treatment full of alders—the whole stand had a much greater total volume than the one in the bare-earth plot, where only a few fast-growing pines remained. A higher number of smaller trees added up to more wood volume than a few large ones.

In time, I'd see that where the alders were thriving, compared to where I sprayed them annually to ensure they and the other plants did not reestablish, they'd continue to add nitrogen to the soil. In fifteen years, there'd be three times more nitrogen than where the alder had been killed. The bare-earth treatment had traded a short-term gain—fleeting water, light, and nutrient increases—for long-term pain, a long-term decline in fixed nitrogen additions. The weeding treatments were robbing Peter to pay Paul.

I RETURNED to finish my grad work in Corvallis and moved in with Don, slipping into the comfort of his small house, and he converted the spare bedroom into my study. We established a pattern of bicycle rides to campus, noon runs along country roads, and meals in the garden. We picked apples and huckleberries, and he made pies. He'd planted tomatoes and squashes to make stews for dinner parties with friends, where his easy, loquacious conversations let me relax in spite of my shyness. I focused on my classes and data, and he worked and cooked and watched the World Series. When he wasn't collecting soil samples, he was running them through instruments, or analyzing data and keeping his supervisor's lab organized. Eight-hour days. He loved the routine. It kept me steady. He made time to show me how to run my samples through a mass spectrometer, figure out the water-holding capacity of soil, and compile my reams of data. Lingering September days led to crisp October weeks. Gusty rains in November turned to snow in December, deep enough to ski through to campus. I read and wrote and learned. He didn't mind my intense focus and became interested in my quest to decode the secrets of pine and alder. On weekends, we'd head to the Cascades, to hike or ski the trails. I finally felt at home in this college town in the Pacific Northwest, and he was glad to be settled in with me. I don't think either of us realized how easy our lives were in those days.

My data warned me of trouble ahead.

It had become clear that removing alder was reducing additions of transformed nitrogen to the soil. Within one year of planting, the effects of alder weeding on soil nitrogen had already become evident in the lowered nitrogen concentrations in the pine needles. On top of that, even though the pines free of alder were growing at faster rates, more than half had died. I worried that in the long haul, in the decades to come, the reduction in soil nitrogen would reduce the growth rates of the open-grown pines that remained. Eventually I would learn that these alderless pines would become so malnourished that they'd be infested by the mountain pine beetle, and most of the rest would die. Three decades later, only 10 percent of the original seedlings planted in the bare-earth treatment would remain.

The weeding proselytizers kept overlooking the fallout from the longer-term loss of nitrogen and eventual decline of the plantations. How could we ignore this? I needed to convince them that alder was necessary for replenishing the soil and, over the longer haul, was complementary, not detrimental, to pine growth. I needed more evidence that alder was a facilitator, not just a competitor. But it could take decades for the impacts of alder removal—the downturn in nitrogen fixation, decomposition, and mineralization—to show up in lost productivity of the forest. I couldn't wait that long. Besides, the seedlings seemed to sense nitrogen depletion almost immediately. The needles of pines in bare earth had less nitrogen than those among the alder after only one year. There had to be a more direct pathway between alder and pine.

I puzzled over *how* the pine seedlings received the nitrogen so rapidly from alder. The conventional thinking was that the transformed nitrogen was stored in alder leaves, which were shed during the waning days of fall and decayed by a food web of bugs. A pyramid of creatures, the bigger ones eating the littler ones. Earthworms, slugs, snails, spiders, beetles, centipedes, springtails, millipedes, enchytraeids, tardigrades, mites, pauropods, copepods, bacteria, protozoa, nematodes, archaea, fungi, viruses, all munching and crunching one another. More than 90 million critters living in each teaspoon of soil. As they ate the leaves, they made smaller and smaller bits of litter. As they consumed the litter, and one another, they excreted excess nitrogen into the soil pores, making a nutritious soup of nitrogen compounds accessible to the pine roots. But in this decomposition and mineralization process, faster-growing plants like grasses could grab the inorganic nitrogen before the pines, and this didn't square with the great amount that ended up in the needles of the pines growing alongside the alders and grasses.

One particularly gruesome study showed that the mycorrhizal fungal threads growing from the tips of roots could invade the stomachs of soil-dwelling springtails feeding on decaying plant litter. The fungal threads sucked the nitrogen straight out of the springtail stomachs and delivered it directly to their plant partners. The springtails, of course, suffered a horrible death. The fungi supplied one-quarter of plant nitrogen simply with the stomach contents of springtails!

I wondered if there might be an even more direct route for nitrogen to transfer from alder to pine that involved the fungi. One that bypassed decomposers like springtails.

I scoured the journals and talked with soil scientists and dropped in at mycology labs. I remembered that the ghost pipe plants of Stryen Creek—the white ones with no chlorophyll—had special monotropoid mycorrhizas that attached to the pines and took photosynthate from them and delivered it directly to the ghost plants. Like Robin Hood.

Then I discovered what I was looking for. After days of searching journals in the university library, I happened upon a new article by a young Swedish researcher, Kristina Arnebrant, who'd just found that a shared mycorrhizal fungal species could link alder with pine, delivering nitrogen directly. I sped through the pages, stunned.

Pine got nitrogen from alder not through the soil at all but *thanks to mycorrhizal fungi*! As though alder were sending vitamins to pine directly, through a pipeline. After mycorrhizal fungi colonized alder roots, the fungal threads grew toward the pine roots and linked the plants.

I figured the nitrogen moved from the rich alder—with its buckets of the stuff—to the poor pine through this linkage, flowing down a concentration gradient.

I dashed out of the stacks and called Robyn from a phone in the foyer. She was back in Nelson for the fall, teaching grade one.

"Just a sec." She yelled at a kid to stop running down the hall while I chattered about pines getting nitrogen from alder thanks to mycorrhizal fungi.

"Wait, wait, wait. How does the fungal pipeline know how to do that? And why would the alder bother to send it in the first place?"

"Oh, well . . ." Leave it to her to stump me. "Maybe the alders have more nitrogen than they need."

"Or the pines give something back to the alders?" she said. "Gotta go!"

I looked at the receiver as it blared a dial tone before racing to Don's office, where he was crunching data. I shouted about finding a cool article that showed alder could connect with pine in a fungal network and send it nitrogen.

"Huh? What? Slow down."

I sank into the chair beside his desk, his computer the size of a television, with a program on the screen spooling through reams of data. I blurted out the details of what I'd read.

"That makes sense," he said. He told me about a new study from California showing the same mycorrhizal fungal species colonizing Garry oak and Douglas fir, and scientists were trying to figure out whether the tree species were linked. And whether nutrients moved between them.

I rummaged in my pack and pulled out some chocolate chip cookies he'd made. They tasted so much better than Ray's had looked, and I was burning energy like a jet as we traded ideas. If a nitrogen-transforming plant like alder could send nitrogen to a tree like pine, the forests might not be as nitrogen limited as we thought.

We chattered about the implications for farms: if legumes passed nitrogen to corn, for instance, we could mix crops and stop having to pollute the soil with fertilizers and herbicides.

My mind was wound up like a clock, pendulum swinging. The direct link between alder and pine, explaining the rapidity with which pine could sense the availability of newly transformed nitrogen in alder, could be the mycorrhizal fungi. The effect of alder removal could immediately be detected by pine because of this link. If I could show how alder sent nitrogen to pine, and how quickly, we wouldn't have to wait a hundred years for the forest to grow in order to show how its productivity was compromised by the removal of alder. The little hand in my brain clock ticked forward and struck midnight.

"Do you think this could stop them from spraying the alders?" I asked.

Don tapped on his keyboard as his calculations finished. "Suzie," he said, "sorry. I doubt it. The forest industry wants fast, cheap wood, and they've perfected growing Douglas fir in forty years instead of hundreds on the Oregon Coast Range. They've been doing this for years. They're making money hand over fist spraying red alder, then adding nitrogen fertilizers." Red alder was a tree, not a shrub like its cousin the Sitka alder, so it was substantially more competitive for light, even though it added more than ten times more nitrogen to the soil. It was number one on the hit list.

A last student padded down the hallway and left for the evening. The measurements I'd taken in my experiments provided only part of the picture. They missed what we could not yet fully see: how the symbiotic bacteria and mycorrhizas in the roots of the alder, and the other invisible creatures in the soil, helped the pine. And they failed to illuminate the bigger picture: that interactions over resources isn't a winner-take-all thing; it's about give-and-take, building more from a little and finding balance over the long term. Don was right; the government and cash-minded companies focused on cheap, quick fixes and the bottom line.

When he saw my shoulders sag, he told me to pull together a solid story and fight them with my data. I perked up; I had lots of experiments where I'd sprayed alder with herbicides showing no improvement in pine growth. But what I really needed was evidence that alder *helped* pine.

"Remember how I took samples at your master's site this summer to see how much nitrogen the alder was fixing, and how much of the mineralized nitrogen was ending up in the pine?" He shut off his computer. "I'll use the data to make some long-term forecasts." If he published his projections, it might help him find work in Canada when we moved back together. He'd calibrate the FORECAST model with my growth and nitrogen data and simulate how different amounts of alder might affect pine growth over the long term. He'd already run the model with red alder and Douglas fir, and Douglas-fir growth declined within a hundred years where red alder was excluded.

"So we *do* have data that will tell us if alder subsidizes pine," I said, excitement rising in my voice again. Sunset was painting the walls in glowing orange.

He picked up his bike helmet so we could head home. The data were only part of the battle. Removing alder might halt nitrogen additions, but so far we only had one-year seedling-growth results. Even with his model runs, I'd still need long-term data to be persuasive.

"Foresters have to *see* results," he said. I'd need that if I hoped to go out into the world with what I was finding.

I clicked on my own helmet, knowing he was right.

"But I'm a lousy speaker," I said. I was terrified of public speaking. "I keep having this nightmare where I spill my slides and have to speak off the top of my head." The only time I'd given a talk cold like that, I'd frozen and almost passed out from embarrassment.

"Yeah, that's why I'll always be a technician," he said. "But you can't hide if you want change."

We biked home. The bigleaf maples hanging over the streets were brilliant gold, and the red oaks were on fire in the cold, clear fall air. We turned onto an empty street, so I sped up to ride side by side with Don. We passed Craftsman bungalows with clutches of students on open porches, reading or engrossed in conversations, and we rode by white multistoried fraternities with expensive cars and throngs of boys playing volleyball and drinking beer. I hadn't seen fraternities or sororities like this at the University of British Columbia where I had earned my undergraduate degree, so this taste of American culture was alluring. I couldn't help but gawk.

We swerved around a dead possum, and I wondered how he had felt about the people who maligned him as he rummaged through composts and barbeques, ignoring his crucial role in eating ticks, slugs, and snails. We stopped at a crossroad, and I asked what the companies might do if science got in the way of earning money.

Don shrugged. "They'll want a policy that protects their earnings. Your story has to be convincing."

He sped up again, and I mulled over how on earth I'd reach the people who could help make changes. I'd learned to deal with conflict by running from it. I was terrible at standing my ground, never mind giving talks.

"Watch out, Suzie!" he yelled, and I slammed on my brakes. A car crossed right in front of us, narrowly missing me.

BY THE TIME I finished my master's, I had gotten more practice giving talks at local forestry conferences. I worked slowly to develop some speaking skills, starting with well-prepared slides and simply presented data, and I practiced my delivery. Then I needed to undo some of these skills and loosen up so I didn't sound boring. I made plenty of mistakes. Like the time I said, "This seedling looks like shit," drawing a complaint from some men that "young women shouldn't swear." But I also earned a compliment from a prominent researcher, who said, "You have a gift for speaking in public." I didn't, but I appreciated his encouragement. I had a long way to go. I had a message but didn't yet know how to convey it in an engaging story.

Don and I moved back to Canada. That fall, when I was twenty-nine and he was thirty-two, we married under the shimmering aspens near Kamloops. I was not in a rush to marry, but we were on a tight timeline if he hoped to stay in Canada. No matter; I was in love and there was no reason not to.

Robyn was my matron of honor and Jean my bridesmaid. Robyn, Jean, and I wore simple skirts and matching blouses. Mine, picked out by Mum, were the creamy white of aspen bark, and Robyn's the shade of the bulrush leaves along the lakeshore. Jean's had little blue flowers the tint of water. Mum and Grannie Winnie wore purple. Mum made cucumber sandwiches and baked the wedding cake, fruitcake drenched in sherry and iced with marzipan. Grannie twisted my hair into a French braid and pinned in baby's-breath flowers. She was quiet as always. After finishing my braid, she straightened my skirt and told me I looked lovely. I knew she was proud that I was as tough as she was, but not too tough. She'd almost died five years earlier from lupus but had rebounded to grow huge gardens—but her tears were never far away as she got older. She managed to stave them off now as she watched me find my place beside Don.

Robyn was with Bill, now her husband, taking low-key candid shots with his camera, and Jean was newly wed too. Dad arrived with Marlene, and they joked with Mum about how Robyn, Jean, and I had all married within three years, and Kelly was sure to follow. "Jeez Louise, with all these weddings!" Marlene offered, great at easing the tension between Mum and Dad. Don's parents came all the way from St. Louis despite bad weather.

Kelly had taken the weekend off and was dressed up in pale blue pants and the navy blue sweater Grannie Winnie had knitted for him, and shoes instead of cowboy boots. It was a tough time of year, one of rounding up cattle and pulling the sprinkler pipes out of fields for the winter, but I loved that he'd been able to come. Tiffany wouldn't have missed this for the world, but her grandmother was ill. Kelly strode up to me with his Mabel Lake smile and the wind in his sails. He had his girl and his farrier business, and he was riding the range. "Congratulations, Suzie," he said in my ear.

Auntie Betty pounded the piano during the ceremony, breaking into "Here Comes the Bride" as the seventeen of us stood in the sun. After Don and I said, "I do," and turned to embrace our families, we all stood still for a moment.

Not long after that, I noticed Kelly off by himself in a grove, hands in pockets, lost in thought. Maybe he was just enjoying a peaceful moment. We'd grown up knowing so well how silence can be soothing. Or deafening. Holding in our feelings, hiding something troubling. Kelly looked up and smiled at me, an assurance that he was all right.

Bill wanted us to pose for photos by the lake. "You're going to get your heels stuck in the mud," he joked, pointing to some humus covered in hoarfrost as I limped in my green-and-violet high heels. "Not to worry," I said, sitting on a log and pulling hiking boots out of my pack.

"Take the pictures from the ankles up, Bill," Kelly said as we took the trail to where Bill captured us laughing in the pale sun under the aspens, with the water starting to freeze along the shore.

My life had knitted together as tightly as the braid running down my back.

7

BAR FIGHT

Blank with fear, I walked to the podium. Under the bright lights, the conference hall was a sea of crew cuts and ball caps. I grabbed the sweaty slide changer. Lavish applause for the previous speaker, a fellow from Monsanto who'd just promoted brushing (weeding) with the herbicide Roundup, dwindled to silence. Still vivid were images of free-to-grow lodgepole pine surrounded by dead aspen, Douglas fir among lifeless birches, spruce free of neighboring huckleberries. Shivering in my blue cotton pants, my white polo shirt drenched, I was glad Barb, my Forest Service research technician of three years, had loaned me her navy blazer. We were both thirty-three, the same not-too-tall, not-too-short size, though our backgrounds were as different as birch and fir—she was the wise mother of three teenagers, while I still had my head in books.

"Thanks for having me," I began. The mike squealed, causing my listeners to wince. Some of the field foresters and policymakers picked up their notebooks, and the young women watched me intently. Others whispered to their neighbors. One guy at the back shouted at me to speak up. The Monsanto fellow hadn't talked at all about whether killing the native plants and reaching free to grow helped the conifers survive better or grow faster.

Barb shot me two thumbs-up as I buttoned the blazer. We were here in the cowboy town of Williams Lake to present my alder research. I'd flown up from Corvallis, where I was working on my doctoral research, digging more deeply into exactly how deciduous and coniferous trees relate to each other. She'd driven her green government pickup truck from Kamloops, three hundred kilometers to the southeast, to meet me at the airport, where I'd instantly spotted her flaming red hair and too-small pink Disney backpack, a hand-me-down from one of her kids.

I'd thrown my arm around her as we strode past large black-and-white photos recording the history of the Williams Lake Stampede, with leathery cowboys risking their lives riding bareback on bulls and broncos, next to pictures of men who'd died too young working the rivers and land for gold, fur, and cattle. She'd warned me of conversations at the Forest Service already questioning the accuracy of our experiments. But I'd jumped at this opportunity because Williams Lake was where Kelly lived, so I'd have a chance to see him. We planned to meet at the pub, and I was hoping Tiffany, whom he'd married a couple of years earlier in a cowboy wedding at the Onward Ranch, would join us. They'd been busy with the rodeo circuit and his farrier work, while I'd established an ambitious reforestation research program for the government and Don had started a forest-ecology consulting business before we'd taken a break to return for our doctorates.

I clicked through the first of my well-prepared slides. Everyone brightened when I showed the sea of leafy green alder followed by the brown stubble left after they'd been cut back. I'd wrenched the quiver from my voice with practice, and I remembered Dad telling me to imagine an audience as a bunch of cabbages. I surveyed the rows of cabbages, rested a glance on Barb, and let out a nervous wheeze while saying, "All the research I'm showing today is published in peer-reviewed articles."

Some of the cabbage heads were nodding at my slides. Barb was cheerful. Don was home, working on his PhD research and baking and riding his bike to classes, having eased back into college life in Oregon after a few years in Kamloops in a log house we'd built in the forest on the outskirts of town. Though we missed our home in the woods, he hadn't mixed easily with the small-town, working-class culture of the pulp-mill crossroads, and he was happy we'd resumed our routine of running and cycling the country roads of Corvallis, him advising me on how to analyze my data and build my confidence.

I exhaled for the next slide. It showed that weeding alder, whether getting rid of all of it or just a little, resulted in zero improvement in pine growth. The weeding treatments meant to help plantations meet free-to-grow regulations—costing the companies millions of dollars—had not increased the performance of the trees. These weeded, free-to-grow pines, in spite of the money spent on them, grew at the same rate as the ones among the alders.

The room went awkwardly silent. Dave, a young, lighthearted forester I'd met at a reforestation workshop, leaned over to his manager, pointing at the slide. They'd been dutifully weeding their plantations, and I was suggesting they didn't need to remove any alder at all. The only treatment in my master's experiment that increased pine growth was the bare-earth apocalypse, where we'd erased every herb, leaf, and grass blade. But these brush-free individuals—the few that hadn't died from frost or sunburn, or become a meal for rodents—had turned into large, gangly trees with big, awkward branches and swollen stems as they feasted on the flood of light, water, and nutrients from the rotting carcasses of their dead neighbors. I took a deep breath to explain.

"There is no benefit to the pines by removing just the alder—you have to kill all of the herbs and pinegrass too if you want fast pine growth," I said, showing slides of the mammoth trees scattered in a heavily weeded patch. The audience murmured at the unusual-looking pines, with their twisted trunks pocked with sores and cankers. These foresters knew these fast-growing trees would have wide growth rings and big knots, much different from the slow-growing trees that naturally regenerated after fire. But their hope was that the cultivated trees would outgrow these defects and still be valuable in half a century, at the time of the next harvest. My data were challenging their hopeful assumptions. They knew too, as well as I did, that they could never achieve these bare-earth conditions in normal operations—the cost of all that manicuring would be out of reach. All they could realistically do was cut back the alder once and leave the understory plants—to no benefit whatsoever, according to my data. But their hands were tied by the free-to-grow policy, which would trigger fines or more costly treatments if their plantations were not weeded of the taller, shrubby alders. I understood the policy was meant to ensure that public forests were left populated with healthy, free-growing trees after the harvests, but in their zeal, the policymakers seemed to have forgotten that the forest is much more than a collection of fast-growing trees. Gunning for fast early growth by weeding out native plants in hopes of future profits was not going to end well. For anyone.

I pointed out that a policy aiming for enormous cash-crop trees wasn't good if it didn't actually produce healthier forests. Focused on my notes, I didn't fully register that the policymakers were now crossing their arms. "You can see in this slide that removing alder somewhat increases the amount of light the pines receive, as you'd expect, as well as water for a week in the middle of the summer, but this is at the cost of lowered nitrogen availability once the piles of dead plants have decomposed. The end result is that there is little net improvement in stand growth after five years," I said, moving on to the data from my weather station, showing how killing all the plants made local climates more extreme—blistering hot during the day and frosty at the soil surface during the night. I was back to stammering as my whirling weathervane, tipping rain bucket, sprawling wires and sensors, and ticking data logger seemed to materialize next to me. Barb, on her way to becoming an award-winning photographer, snapped a picture as an attempt to fortify me.

A hand was waving, and I motioned in its direction. "That's your special research site, but what about the real world?" asked a forester. The men around him nodded.

"Great question," I answered, excitement rising. "I've been following the responses of planted trees to regular brushing operations, where the companies have cut back the alder but left the grasses and herbs in place, and compared them to control plots left untreated. We've found the same result over and over. These practices do ensure the trees are free to grow—taller than the remaining plants. But whether sprayed with herbicide or cut with brush saws, whether on dry or wet sites, down south or up north, or whether the crop is pine or spruce, stand growth doesn't improve even though they can be declared free to grow sooner. What bothers me is that half the free-to-grow pines now have some infection or injury that will eventually kill or maim them."

One of the head policymakers in the Forest Service scowled. He'd had my colleagues review the paper I'd written on this study to look for flaws even after I'd published it in a peer-reviewed journal. He was known as the Reverend because of the gospel he preached about policies he'd helped write. Ones shaping the species composition and health of the entire forest landscape. Beside him was Joe, the vegetation management forester in the same Forest Service office where Barb and I worked, and where she'd overheard discussions about the reliability of our experiments. The Reverend and Joe together suddenly felt dangerous.

Barb nodded at me, a signal to stay on track. I added another piece of the story, showing how Don's model had forecast that productivity of century-old pine forests would dwindle by half where alders were no longer present to add nitrogen back to the soil. And that the vitality of the forest would slide further and further with removal of alder at each successive cutting cycle. The model showed how pines needed alder neighbors to supply them with new nitrogen so they could grow to make healthy forests, especially when the nitrogen capital was depleted right after a disturbance such as logging or fire.

A young woman's hand shot up, and she didn't wait for a gesture from me before asking, "Then why are we spending so much to spray alder when it doesn't improve the performance of our plantations and might even make them worse off?"

There was shifting in seats and whispering. The muscles in the back of my neck tightened, but I pushed on, saying directly to her, "We should take a closer look at the free-to-grow policy to see if these costs are justified. I am worried about the future health of our plantations." I wished Alan were here, and I clung to the thought of him and his support for my research and me. He would have helped address these questions.

The Reverend said something to Joe that made them both laugh. They were no longer cabbages. Emboldened, Joe stuck up his hand and declared my results premature before asking, "Shouldn't we take a more cautious approach and wait for longer-term data?"

His tone was mild, but his position was clear. Early on, Joe had supported my work, then changed his mind as the results started taking shape. He was out to make a name for himself, and disagreeing with the policies of the higher-ups wouldn't get him far. I cautioned myself not to show weakness. If I agreed that my work was still incomplete, I'd be dismissed and nothing would change. Barb shifted forward in a way that encouraged me to address his question head-on. I leaned into the mike. She veered her pile of red hair in Joe's direction, throwing him a dagger look. Surprising myself, I calmly argued that it would be great to have long-term results in hand, but these studies were already harbingers of the future. Lack of growth increases at this young age weren't likely to transform into big improvements down the road. We shouldn't be counting on improved productivity. I went on, "These brushing treatments to meet free to grow appear to be putting our plantations at risk of high early mortality losses and lower long-term growth. The more cautious approach would be to let these plantations grow up with the native plant communities intact while we focus on other weaknesses in the silviculture plan, such as when to plant, what to plant, and how to prepare the sites."

A few guys got up to leave. One in the front row started talking loudly to his buddy. I tried to signal that he was interrupting me, but he kept on, making me try harder, as I'd done as a kid playing street hockey. Dave, who'd always been open to discussing new practices with me, frowned at the interrupter.

I suppressed the urge to stop and ask the interrupter whether he had a question. But I had shrunk inside and didn't want to create a scene. What would Grannie Winnie do? She'd stubbornly, quietly move on. My hands were shaking, but I advanced to the next slide and continued with my surfeit of experiments in other plant communities. One hundred and thirty, to be exact. All replicated, randomized, and with solid controls, all coming to similar conclusions.

Cutting or spraying willows did not improve spruce growth or survival.

Cutting, spraying, or grazing fireweed with sheep did not improve performance of either spruce or lodgepole pine.

Neither cutting nor grazing thimbleberry helped spruce.

Cutting aspen did not increase the girth of pines.

Whether we sprayed, cut, or grazed the rhododendron, false azalea, and huckleberry communities in the high-elevation plantations, the growth of spruce didn't budge. I thought back to Robyn spraying the rhododendrons, and how we'd suspected even then it was a waste of time.

In these high-elevation forests, a lot of cash was spent to grow seed-

Old-growth forest with ancient western red cedar Mother Trees in the overstory and western hemlock, amabilis fir, huckleberries, and salmonberry in the understory. Western red cedar is known as the tree of life to the Aboriginal people of the West Coast of North America, for whom it is of great spiritual, cultural, medicinal, and ecological significance. The wood of the species is important for making totem poles, house planks, dugout canoes, paddles, and bentwood boxes, and the bark and cambium are used for making baskets, clothing, rope, and hats. British Columbia recognizes western red cedar as its official tree.

Suillus lakei, or painted *Suillus*, or pancake mushroom. This species is ectomycorrhizal and only grows in association with Douglas fir. The mushrooms are edible, although not highly prized, and are eaten in soups and stews. The cones beneath the mushroom caps are Douglas fir. The plants in the foreground are creeping raspberry and bunchberry. The Haida mix raspberries with bog cranberries and dry them.

Douglas-fir Mother Tree, approximately five hundred years old. The thick, deeply grooved bark protects the tree against fire, and the large branches provide habitat for birds and wildlife such as winter wrens and crossbills, squirrels and shrews. The Aboriginal people use the wood for pit fires and making fishing hooks, and the boughs for floor coverings in lodges and sweathouses.

Western red cedar Grandmother Tree, approximately a thousand years old. The vertical crevices suggest the bark has been stripped by First Nations people. The inner bark is separated from the outer bark and used for making cedar baskets and mats, clothing and rope. Before the harvest, the people place their hands on the trunk to pray and ask permission for the harvest, and in so doing they develop strong connections with the tree. Strips up to one-third of the circumference and thirty feet long are harvested, leaving a shallow scar narrow enough to heal over.

Mycena, or bonnet mushrooms. The *Mycena*s are saprotrophic, or decay mushrooms, and are generally not eaten.

Amanita muscaria, or fly agaric. The species forms ectomycorrhizae with many trees, including Douglas fir and paper birch, as well as pines, oaks, and spruces. The mushrooms can be poisonous and are psychoactive.

Sitka spruce Mother Tree on Haida Gwaii. The hemlock saplings in its understory are regenerating on decomposing nurse logs, which protect the new regeneration from predators, pathogens, and drought.

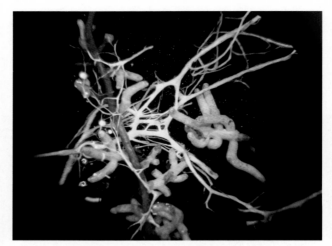

Ectomycorrhizal fungal root tips and rhizomorphs

Suillus spectabilis, also known as the bolete filter, is an ectomycorrhizal fungus. The white fungus mantle is enveloping a tree root tip and forming a pinnate structure. The mantle serves to protect the root tip against damage or pathogens and is the foundation from which fungal mycelia emanate to explore the soil for nutrients. The mushrooms are edible but sour and pungent.

Pointing to the white fungal mycelium that weaves through the forest floor

Western red cedar in a Pacific coastal rain forest, probably five hundred years old. This tree is the cornerstone of Aboriginal culture on the West Coast of North America. Although the tree is used for many important cultural items—clothing, tools, and medicines—few were felled before European contact. Instead, fallen trees were gathered or boards were split from living trees, where yew or antler were wedged into the grain.

Western red cedar Mother Tree with her offspring. Cedar reproduces by seed but also by layering, where the branch of the Mother Tree swoops down to touch the earth then develops roots at the point of contact. Once the branch is firmly rooted, the sapling severs from the parent and becomes an individual tree. The maple to the right of this tree is a common tree associate of red cedar, as they both prosper in rich, wet soils and are joined together in an arbuscular mycorrhizal network.

lings in open spaces where they didn't naturally exist. True enough that
20 percent more seedlings survived where the non-cash-crop shrubs
were weeded than where they were left untouched—*but only for the
short term*. In the same subalpine environments, spraying ferns into
pincushions did not improve the long-term survival rate of spruce,
but the short-term height growth of the prickly seedlings was a quarter more than where the ferns were left alive. These minimal, tempo-
rary yields were enough to satisfy the policymakers.

"I've given a lot of thought to why we don't usually find improvements in tree survival or growth when we weed out the native plants,
even though the trees are now free to grow," I said. "And I've thought
about why many of the free-to-grow trees are becoming infested with
insects or pathogens—and are worse off. For one, I think we're overestimating how competitive these native plants are with the conifers.
On most of the sites, the native plants do not grow back so densely
that they hinder the trees. I also suspect the plants are protecting the
trees against blights and severe weather. We should change our focus
from weed-free trees in hopes of short-term growth gains and instead
consider what makes the whole forest healthier over the long term."

I thought of some friends who'd plucked their eyebrows so thoroughly that they no longer grew back. Not an analogy I could share
at this conference.

I explained that getting the kind of harvests we wanted was making
us treat the new forests like farm fields. And the free-to-grow rules
were being applied to any and all landscapes. Lots of money spread
over lots of terrains, usually with reduced plant diversity.

Barb and I called it "the fast-food approach to forestry." The same
broad brush applied to all sorts of forest ecosystems was like delivering
identical burgers to all cultures, whether in New York or New Delhi.
A man in a yellow Finning ball cap in the third row pulled out a bag
of carrots and started loudly crunching them. It was almost time for
the coffee break.

"We're barking up the wrong tree with brushing," I said. A few foresters laughed. Barb guffawed, as she always did at my bad jokes, but
the rest were stony faced.

A policymaker raised his hand. "You've cherry-picked studying the
plants we aren't as worried about. We already know these ones aren't a

big deal." The Reverend was nodding, even though he and the speaker knew these plants were popular targets of their policy. "What about the more competitive ones like pinegrass and paper birch?"

"Good point," I said. "Pinegrass is good at soaking up water and nutrients from the soil, but we've found that spraying it or scraping it away with an excavator shovel increases pine seedling survival and growth only by about 20 percent. And there are unexpected side effects—the treatments also compact the soil and reduce nutrient content. They increase erosion and reduce the diversity of mycorrhizal fungi."

"We're putting excavators on all our pinegrass sites. Are you saying it's not worth it?" a young woman asked. Relaxing a little, I scanned the room and found her eager face, her auburn hair swept into a topknot. She didn't seem bothered by the heavy silence that had fallen around her. The Reverend turned to see who in the world would ask this question.

"Well, we need to better understand what we are winning and what we are losing at the same time," I said. "Maybe there are better ways to improve the plantations than raking away the forest floor. It doesn't bode well for long-term health when we remove organic matter and compact the soil. We need better data before applying these treatments over the whole landscape."

"Suzanne, what about the birch?" came from the back of the room. "This is what the free-to-grow policy is really meant for." It was a scientist from Victoria also trying to get to the bottom of how birch and aspen were hindering or helping the conifers. Like me, he was interested in the far-reaching ecological consequences of brushing, but his position was more embedded in policymaking.

At last a segue to my results for the plant communities containing paper birch. "You're right. Cutting, spraying, or girdling birch with chains increases the girth of Douglas fir, sometimes by one and a half times," I said, flicking to a histogram of fir-growth responses to the different treatments. A shaft of sunlight briefly streaked across the room when someone leaned against a blind, and the cabbages tilted forward. I wished I could run outside into the clear, free air. I'd wanted to talk about birch, but it would open a can of worms.

Joe nodded to the Reverend and pointed at the slide, finally seeing what he wanted.

"But we have to be careful, because the more birch removed, the more firs die from root disease," I said. "Cutting and girdling stresses the birch and makes them vulnerable to root infections. As soon as we cut the birch, the infection overwhelms their roots and spreads to the roots of the firs, causing seven times the infection rates of untreated stands. I'm worried that we're trading increased early growth for lower survival in the long run."

A pathologist interjected that since my studies weren't showing how disease behaved over whole forests, I should be cautious. The pathogenic fungi grew in distinct patches, and without knowing exactly where these lay belowground, the random allocation of my two test plots—one treated and one controlled—could accidently be linked to a virulent patch, or not. He thought I might have found these results by chance, by the sheer luck of placement. In other words, I needed to be studying the behavioral responses of the pathogen over larger areas. We'd had this discussion privately, and we'd agreed that since I'd repeated the experiments on so many sites there was validity to my findings, so I was frustrated he was raising these uncertainties at this moment.

"Yes, that's true," I said as blandly as possible, "but this experiment was replicated fifteen times, so I have confidence in the results." The cabbage heads swerved toward the pathologist for the final word, and he shook his head ever so slightly to signal his authority over the topic. A loud bite from the carrot-crunching fellow seemed to confirm this.

I finished my presentation to a smattering of applause. The field foresters were grateful that the evidence matched some of what they were seeing in the bush, but the policymakers were muttering. They continued to question me in that familiar way, explaining that the plantations would become brush fields if "weeds" like birch were left unchecked. They needed longer-term data, if they better fit their vision. They certainly weren't going to shift policy because of my experiments. The crowd dispersed for the break.

People stood in cliques, sipping metallic-tasting coffee and eating muffins. I dropped my slide tray, and slides flew in all directions. A

young man rushed over to help. The rest glanced over and returned to
their conversations. I shakily poured some coffee, not wanting any but
at the same time needing to hang around for feedback. Some foresters
said, "Good talk." Dave said my results made sense, but they'd have
to keep weeding the brush anyway because, well, those were the rules.
Policymakers stayed deep in conversation. None seemed interested in
approaching me, and besides, the Reverend was holding court. Barb
joined me, knowing how difficult—even humiliating—it was waiting
for acknowledgment. I was awful at small talk at the best of times, and
now I was a jumble inside. Finally she grabbed me to head outside.
Breezes washed over us, and a whiskey jack flew past.

"Those bastards!" she said. "At least they could have thanked you
for the work you've done to try to understand what the hell we're
doing to our forests."

I was completely drained. The same way Robyn and I felt when we
removed our drenched pesticide suits after spraying alders. Every cell
in my body depleted. Doing something we hated yet loved. Barb took
photos of the old spruces and quaking aspens at the edge of the parking lot, young spruces seeded into the understory. The young woman
with the auburn hair in a topknot stopped to thank me. The policies
weren't going to alter overnight, but if some of this resonated with
other concerned foresters, perhaps change was possible.

THE DIM OVERLANDER PUB smelled like stale beer and cow shit.
Kelly, his cowboy hat and boots scuffed, was horse-trading at the bar
with Lloyd, a grizzled rancher. Their bony elbows leaned on the worn
waxed slab of the bar, and their bowed legs were spread wide, claiming
their territory. I tried to catch Kelly's attention, but he was relishing
his banter with Lloyd—a language that suited his natural cadences,
his long pauses. I was close enough to hear them negotiate over some
Appaloosa stud, but I sensed this deal would take more hard-fisted
slow talks. Kelly was ignoring me as long as he could, like when we
were kids and I wanted his attention.

This bugged me more than usual after the day's dismissiveness—the
sense of exclusion. Barb nudged me to check out a half-full brass spittoon in the corner. Dressed in T-shirts and shorts, obviously from out

of town, we were catching the eyes of the cowboys. One guy stared at the conference jacket I'd slung over my shirt and whispered something funny to his pal. I didn't much care. I wanted to see Kelly, because we hadn't been able to meet up over the past year. Agitation rose in my throat that he couldn't rip himself away from his horse-trading to come over and say hi. Tiffany wasn't here, and I missed how she'd be pointing out to Kelly that he should be more attuned. Sensing my impatience, he signaled he'd be another two minutes.

I was ready to blow the joint five minutes later, but Barb bought us a jug of beer, sat herself at a corner table, and gestured at me. When Kelly's deal with Lloyd came to its predictable stalemate, he sauntered over, carrying a jug. Lloyd had a lot of cash and was buying. A grin spread across Kelly's wide jaw, and my upset vanished. It was great to see him. We got to the serious business of drinking. It had been a rough day.

"Seen Uncle Wayne lately?" I asked.

"Yeah, the son of a bitch got me a job herding cattle for the Caribou Cattle Company." He spat a stream of rich brown tobacco that made an expert splash landing in the spittoon. Barb's eyes widened in awe, which made me proud of my little brother. He was exotic. Striking. One of a kind. I had missed Kelly and his dogged pursuit of this relict life. A bull-riding, snuff-chewing, calf-birthing, blacksmithing cowboy reincarnated from the past.

"You living on the ranch?"

"Yeah, Tiff and I are in the bunkhouse on the Onward, near the Mission. You know, the Indian Residential School the pedophile priests ran." He looked at his feet in disgust over what those bastards had done to the kids. This part of Canada's history was shameful. Kelly and I knew kids who'd been at the school and had seen firsthand how it destroyed plenty of them. Some had escaped, like our friend Clarence, now a traditional cedar-totem carver on Haida Gwaii.

Friends of Kelly walked into the bar, shouted hello, and yelled, "Can you come shoe my horse tomorrow?" He flicked his ham-sized hand to tell them yes, sure.

"Where's Tiffany?" I asked.

"She's got morning sickness." Kelly couldn't hide his mountain-high pride.

"Whoa, that's crazy! Congratulations!" I jumped up and gave him a high five. Hugs didn't exist in our family, but smiles and hand gestures did the trick.

Lloyd wandered over from where he'd been jawing with another cowboy and filled our empty mugs for a toast. I sloshed mine a bit too hard, so Lloyd topped off my glass. Kelly was starting to drawl like a Texan.

"How'd your meeting go?" He was slurring a little.

Barb jumped in, "Those assholes didn't like hearing from a woman."

"They didn't believe me," I said quietly. I especially hated it when Joe had leaned over to the Reverend when I showed the nitrogen and neutron probe results. My body tensed recalling it, my usual bracing for a quick escape. There weren't many emotional topics my family spoke of easily. I glanced at Kelly's friends across the bar.

"Those forestry guys don't know how to work around cattle, either. It's bloody hard work, and they expect us to herd our cows out of their plantations fast, at the drop of a hat. I gotta be on my horse at dawn," said Kelly.

I laughed. The room was starting to float. I staggered to the bathroom, thinking of how Kelly and I used to flee the tension between our parents by riding our bikes in the forest, lassoing old stumps as though they were calves. I returned to another mug of beer.

"You can get the cows moving, though," Kelly said. He was as drunk as I was. "If you manage them like women."

I stared at his swimming eyes, not sure I'd heard right. I was always so taken by surprise by comments that offended me that I would try to change their meaning in my head. Pretend that something said really hadn't been. Or I'd revise what was said into something softer, often agreeing when I didn't. This time, I was too drunk to twist it around. Barb was sitting straighter, even though I was sure she was equally shit-faced.

"What do you mean?" My flushed cheeks were burning. Way on the other side of the bar, the jukebox had flipped to Willie Nelson's gravelly voice, singing about mamas and babies. I wished I could turn off the conversation, get rescued from whatever was coming. Barb's chair made a scraping sound as she pushed her hands against the table

Kelly calf roping at a local rodeo, early 1990s. With the calf lassoed, he will make a running dismount, catch the calf, and then tie three of its legs together.

in an attempt to get up, maybe scrambling to figure out what to do, how to interrupt, anything to throw a wet blanket over me. Lloyd grinned in amusement from over at the bar, pointing at us so the bartender would send over more jugs. Egging us on with more booze.

"The cows are the center of the herd. Their only job is to feed their calves." Kelly was circling his hand above his head as though lassoing one of them.

"Women don't just feed babies. You're kidding me, right?" I was too hammered to tamp down the strain in my voice, all of the world's injustices in my throat. Sober, I would have ignored this. Realized that Kelly wasn't saying these things to offend me. He was winding down from a long week on horseback. But at this moment, I wanted to god-damn well throttle him.

He sloshed onward. "It's the steers that count. They control the cows."

"Are you serious?" My amygdala firmly hijacked my prefrontal cortex.

The bile in my stomach surged, and I pushed away my jug of beer. Barb picked it up and made her way to the bar gingerly, returning it as if handing over a wayward truant.

Kelly mumbled something else about his fucking cows.

"We can do what we goddamned well please. We can even be the bloody prime minister if we want!" I swerved in my seat, my blurry image drifting across the mirror behind him. I sure didn't look like a prime minister. What the hell was I saying?

I didn't hear what Kelly said next, beyond, "Huh?" Everything was so fuzzy I could barely see his face across the table. Barb said we needed to go. I staggered to my feet and tried to put on my jacket.

"Fuck you!" I shouted, stomping off with one arm in a sleeve and the rest of my jacket dragging. Cowboys swilling whiskey at the bar turned their heads, and one let out a low whistle.

Kelly roared something at my disappearing back as I stumbled out of the Overlander, jukebox wailing.

I flew to Corvallis with the worst hangover of my life. My head ached and my lips burned. After walking in the door at home, I flopped on the couch with a wet cloth over my eyes. Don hugged me and told me I would be fine and that Kelly would get over it.

BUT INSTEAD I ended up in a cold war with my brother as well as the policy guys. The irony of the bar fight is that it slammed up against the very question I had been pursuing in my doctoral dissertation about collaboration in nature. Are forests structured mainly by competition, or is cooperation as or even more important?

We emphasize domination and competition in the management of trees in forests. And crops in agricultural fields. And stock animals on farms. We emphasize factions instead of coalitions. In forestry, the theory of dominance is put into practice through weeding, spacing, thinning, and other methods that promote growth of the prized individuals. In agriculture, it provides the rationale for multimillion-dollar pesticide, fertilizer, and genetic programs to promote single high-yield crops instead of diverse fields.

Speaking out about the stewardship of the land felt like the chief purpose of my life. But I'd already tried and failed abysmally to con-

nect with the people in charge. I had grave doubts about how to proceed, given how easily I felt dismissed and how badly I'd handled the
bar fight.

Meanwhile the clear-cuts around the province were growing like a
cancer, and foresters were killing "weeds" as though we were at war.
Activists were rising up, chaining themselves to trees. Clayoquot
Sound was the scene of big protests against clear-cutting, but I con-
cluded that I might be of more use by focusing on my research.

That summer, I went home to the forests where I'd grown up. I
sent Kelly a postcard in apology, but he never wrote back. Mum said
that Tiffany's pregnancy was going fine, but it hurt that he wasn't
communicating with me. Soon I'd be an aunt, and I wanted to be
involved. I decided to wait for him to come around in his own time. I
wouldn't crowd him. We used to spend hours in silence as kids when
making forts out of the birches fallen under the shade of the firs, and
he needed long open spaces to be himself. We would be okay.

With my head down, though, I wondered why Kelly was taking so
long to respond. Why was it always so much work to stay connected?
To be a family?

8

RADIOACTIVE

Barb and I hauled forty waist-high shade tents out of her truck bed. "Jeezus, these are heavy," she said. Each weighed about ten pounds and was made of shade cloth sown into a cone and fitted over a tripod of rebar. The yellow handkerchief covering her red curls was coated in mosquitoes—at their zenith in mid-June—and her muscled arms glistened with sunscreen and bug dope. She was a steady hand on the outside, warm as a beating heart on the inside. We'd driven over the mountain from Vavenby, a railroad town eighty kilometers south of Blue River, to a clear-cut at the north end of Adams Lake to set up the main field experiment for my doctoral research. It was one of six experiments but by far the most important.

Young birches had already sprouted from cut stumps, and some had grown from seed sprinkled in from the neighboring woods and were taller and growing twice as fast as the conifers we'd planted a year back. I wanted to know whether these birches were simply competitors—reducing the resources Douglas fir needed for survival and growth—or whether they were also collaborators, enhancing the conditions under which the whole forest could thrive. And if the leafy native plants did collaborate with their needle-leafed neighbors, I wanted to know how. To help answer these questions, I was testing whether paper birch donated resources at the same time that it shaded and depressed fir's ability to make food through photosynthesis. As the birches intercepted light for their own sugar production, did they make up for the reduced photosynthetic rate of understory Douglas firs by sharing their riches? My investigation would help me figure out how in the world fir could survive and even prosper in spite of living among birch neighbors considered by foresters to be strong, unwanted competitors. And if birch did spread this bounty—the large amount of sugar it was able to produce in full light—maybe it was delivered to the shaded Douglas fir through a belowground pathway, mycorrhizal fungi linking the two species together. Birch cooperating with fir for the greater health of the community.

"I'm not a very good seamstress," I mumbled as we tightened the wires fastening the rough fabric to the tripod legs.

"But they're built like brick shithouses," Barb said, admiring the collection standing shoulder to shoulder like the Pyramids of Egypt. "Nothing will blow these over." She wasn't going to let me feel sorry for myself for long.

They just had to last a month. That would be long enough to depress the photosynthetic rate and sugar production of the firs. The thick green tents would cut out 95 percent of the light, while the thin black ones would cut it in half. It had been two months since the bar fight, and Kelly still hadn't been in touch, but Barb had assured me that he would reach out in his own time.

Barb and I lugged the tents through the clear-cut to the plantation of test trees, wrangling the pyramids over logs and through clusters of falsebox and fireweed. Our pockets were stuffed with measuring tapes, calipers, and notepads that we used for checking the pulse of the seedlings as we tented them. From a paper bag filled with sixty bits of paper marked with "0," "50," or "95," I pulled one out— a rabbit from a hat—to allocate the shading treatment randomly. I did this to avoid bias in the response of fir that could be caused by something other than shade, something unbeknownst to me, like an underground spring. The shred read "95." I placed a cone covered in heavy green cloth over the fir, encasing it in deep shade as the rebar legs landed inside the foot-deep band of sheet metal I'd sunk the year before to contain the entwining roots of each designated triplet—one birch seedling I'd planted close to one fir and one cedar. I wiggled the rim—the sheet metal was firmly in the ground—and I pushed down on the top of the cone until the rebar legs were solidly in the earth. I pulled a wrinkled map from the pocket of my rust-stained jeans. I loved maps; they led to adventure, discovery. This one showed where we had positioned the sixty triplets, scattered across an area the size of an Olympic swimming pool.

My plan was to cover one-third of the Douglas firs with the heavy green tents and another third with the light black tents, while leaving the remaining third in full sun. This would create a gradient of light reaching the firs, from very little in the deep shade to the most possible in full sun. I was emulating the variety of shady and sunny spots that young fir experience when growing in the shifting shadows of naturally occurring overtopping birch saplings.

But unlike naturally growing birches, which normally seed in or sprout from cut trees immediately after clear-cutting, and thereby have a height advantage over planted conifers, my birches were the same height as my planted firs. They cast no shade at all in my experiment, so I needed to create it artificially, with these tents. Unlike in nature, however, the tents would only provide shade and not simultaneously change the availability of soil water or nutrients. They would help me pinpoint the effect of shading as an isolated factor unaffected by other unseen relationships.

Choking on a mosquito, Barb pulled out her bug hat—a wide-brimmed fedora with a veil of fine netting—and remarked that I was lucky the Forest Service was letting me study whether birch cooperates with fir.

"I tucked this in with the other experiments," I said, smiling. I was getting good at hiding the controversial ones among the mainstream studies when I applied for grants.

I'd been intrigued by the possibility that birch and fir might trade sugar through mycorrhizal fungi ever since I'd read of a discovery in the early 1980s by Sir David Read, a professor at the University of Sheffield, and his students, who'd found that a pine seedling had transmitted carbon belowground to another pine. He had established pines side by side in transparent root boxes in the lab. He'd inoculated the seedling roots with mycorrhizal fungi to link them in a belowground fungal network, then tagged the photosynthetic sugars produced by one of the pines—the donor—with radioactive carbon. To do this, he'd sealed the shoots of the pines in transparent boxes and replaced the naturally occurring carbon dioxide in the air of one of the seedlings with radioactive carbon dioxide. He allowed the pine to absorb it and convert it to radioactive sugars through photosynthesis over a few days. Then he'd placed photographic film over the side of the root box in hopes of recording any radioactive particles that might be transmitting through the network from the donor pine to the receiver pine. Upon developing the film, he saw the path the charged particles had taken as they moved from pine to pine. They had traveled through the underground fungal network.

I wondered if this could be detected outside the lab, in real forests. Sugar might transmit from the root of one tree to another. If so, maybe the added radioactive carbon-14 traveled only between trees of the same species—as Sir David found—but what if it got conveyed between *different* species of trees mixed together, as they're often found in nature?

If carbon did transmit between tree species, this would present an evolutionary paradox, since trees are known to evolve by competing, not cooperating. On the other hand, my theory was entirely plausible to me, because it made sense that they would have selfish interests in keeping their community thriving, so that they could get their needs met too. I worried about what the Forest Service folks would think, but I couldn't let go of this possibility. The pine donor in Sir David's experiment sent carbon to a receiver seedling, and even more when the receiver was shaded, but he did not know if the receiver sent any carbon back. If the donor received as much in return from its neighbor as it gave, that signaled a balanced transaction, with no gain by either individual. Sir David's experiment could never reveal this, because he'd labeled only one of his seedlings with radioactive carbon and hadn't added a tracer to see if the receiver returned as much in the other direction. But if one did gain more, was it enough to help it grow? If so, this might challenge the prevailing theory that cooperation is of lesser importance than competition in evolution and ecology.

I started visualizing birch and fir along the shores of Mabel Lake, connected belowground by mycorrhizal fungi in the same way as the lab pines, sending messages back and forth through the hyphal links. Like having a conversation over the World Wide Web, invented in 1989, only a few years previously. But instead of words, I dreamt that the messages were made of carbon. I thought back to my plant physiology classes, imagining a birch leaf photosynthesizing—converting light energy to chemical energy (sugar) by combining carbon dioxide from the air with water from the soil. Because of their ability to photosynthesize, the leaves were the source of chemical energy, the engines of life. The sugar—carbon rings bonded with hydrogen and oxygen—would accumulate in the cells of the leaves and the sap then load into the leaf veins like blood being pumped into arteries. From the leaves, the sugar would travel into the conducting cells of the phloem—the blanket of tissue encircling the birch trunk under the bark and forming a pathway from leaves to the root tips. Once the sweet sap was in the uppermost sieve cells of the phloem, an osmotic gradient would develop between them and adjacent phloem cells. Water taken up by the roots from the soil would travel up the xylem—the innermost vascular tissue linking roots to foliage—and be loaded into the top sieve cells of the phloem by osmosis, diluting the solution to balance the concentration with the interlinking sieve cells. The increase in pressure in the cells—*turgor pressure*—would force the photosynthate downward through the smooth chain of sieve cells, eventually reaching the roots. The roots, as with the aboveground parts of the tree such as buds and seeds, need energy and are *sinks* for this sugary burst. (While leaves are the *source* of photosynthate, roots are sinks.) The root cells would quickly metabolize the sugar and move some into adjacent root cells, taking water with it and relieving their turgor pressure. The exodus of sugar water from one root cell to another plays its own role in the source-sink gradient as the solution keeps flowing from roots to leaves, and then from the top of the tree to the bottom, a process that scientists call *pressure flow*. It's like blood pumping from our bone marrow (our source) to our vessels then cells (our sinks) to meet our needs for oxygen. As long as leaves synthesize sugars through photosynthesis, enhancing the source strength, and as long as roots keep metabolizing the transported sugars to make more root tissues, enhancing the sink strength, the sugar solution keeps moving by pressure flow down the source-sink gradient from leaves to roots.

Barb and I carried more tents farther down the slope to the remaining triplets. I was taking a risk with this experiment, since underground networks were not yet known to form in forests, never mind between trees of different species. Even more far-fetched was the idea that the networks might serve as avenues for collaboration and trade routes for sugar. But I'd absorbed the merits of synergy from growing up in the forest. From hiking the heavily treed slopes up Simard Mountain. From climbing trees and building shelters with Kelly.

The sugar train in my imagination didn't stop at the roots. I'd read that the photosynthate was unloaded from the root tips into the mycorrhizal fungal partners, like freight unloaded off boxcars onto trucks. The fungal cells engulfing the root cells and extending from there as threads into the soil would be flooded with the sugar. Water brought up from the soil would rush into the receiving fungal cells to balance the sugar concentration with that of the neighboring fungal cells, just as it did in the leaves and phloem. The increasing pressure from the influx of water would force the sugary solution to spread through the threads of fungal cells enveloping the roots and then out through the hyphae emanating into the soil, like water flowing from a tap through a suite of conjoined hoses. Some of the sugars would fan out to help grow more hyphae through the soil, which would also help collect more water and nutrients to bring back to the roots.

My plan was to label paper birch with the radioactive isotope carbon-14 so I could follow the photosynthate traveling to Douglas fir, and at the same time I'd label Douglas fir with the stable isotope carbon-13 to trace photosynthate moving to paper birch. That way I could tell not only if carbon was passing from birch to fir but also distinguish if it was moving in the opposite direction, fir to birch, like trucks on a two-lane highway. By measuring how much of each isotope ended up in each seedling, I could also calculate whether birch gave more to fir than it got in return. Then I would know if trees were in a more sophisticated tango than just a competition for light. I'd discover if my intuition was right—that trees are tightly attuned, shifting their behaviors according to the functioning of their community.

MY EXCITEMENT at checking the seedlings a week later was refreshing after my continuing worry over Kelly. They'd grown vigorously, from ankle-high to knee-high. As Barb and I moved from triplet to triplet, the seedlings greeted us with fragrant bouquets and soft dapples. The little trees were firmly alive. "You look like you'll tell me some secrets," I murmured as I tugged at a fir with a stout stem. Its bottlebrush needles were already touching the soft serrated leaves of its birch neighbor. The cedars were glowing where the birch cast a cool shadow, protecting their delicate chloroplasts from the high sun. Where the birch leaves could not reach, the cedars were tanned red to prevent damage to their chlorophyll. The seedlings in the threesome were so close they seemed bound in a common story—with some type of beginning, middle, and end.

Barb asked why I'd included cedar next to the birch and fir.

Cedar can't form mycorrhizal fungal partnerships with the birch and fir for the simple reason that it forms arbuscular mycorrhizas, not ectomycorrhizas like the other two. If cedar roots acquire any of the sugars fixed by fir or birch, they would have picked it up after it was leaked from their roots into the soil. I'd planted cedar as a control, to tell me how much carbon was leaking into the soil versus how much might be transmitting through the ectomycorrhizal network linking birch and fir.

Using a portable infrared gas analyzer, a contraption the size of a car battery with a see-through barrel-shaped chamber, Barb and I checked that the shade tents were doing their job of depressing the photosynthetic rates of the fir seedlings. I squeezed open the set of jaws and clamped the chamber over the needles of an untented fir. When trapped inside, the fir needles continued to photosynthesize, but instead of the gases floating about in the air, they were forced to run through the little machine. In other words, the gas analyzer measures the rate of photosynthesis.

The sun shone through the clear plastic of the chamber, and the needle swung across the meter. The fir needles were hungrily absorbing the carbon dioxide in the chamber, and the machine was telling me the fir was photosynthesizing at the highest possible rate. Barb jotted down the number, and we moved to the next triplet where the fir was heavily shaded, receiving only 5 percent of the light. After I wriggled the chamber under the shade tent and clamped it over the fir's needles, I breathed a sigh of relief. My shade tents were working. The deeply shaded fir seedling was photosynthesizing at only a quarter of the rate of the fir seedling in the full sun. It was also a relief to see that the tents weren't affecting the air temperature—they were porous enough that air flowed freely—which could have affected the photosynthetic rates. We ran to the next tree, this one under a black tent. The partially shaded seedling was photosynthesizing at a rate in between.

Moving from fir tree to fir tree, we confirmed the pattern. Then we tested the birch. The fully illuminated birches were photosynthesizing at double the rate of the fir seedlings in full sunlight. Eight times the rate of the firs in the deep shade of the green tents, confirming there was a steep source-sink gradient between them. If the two trees were linked by a mycorrhizal network, and if carbon did flow through the connecting hyphae along a source-sink gradient as Sir David thought, then the surplus of photosynthetic sugars in birch leaves should flow into the roots of fir. From source birch leaves to sink fir roots. I was flushed with excitement as I scanned the columns of data. The more shade the tents cast, the steeper the source-sink gradient from birch to fir.

At the end of the day, we loaded the gas analyzer back in the truck. Sitting on the tailgate, I checked to make sure we hadn't forgotten anything. Barb had marked down the concentrations of carbon dioxide, water, and oxygen, the amount of light shining on the needles, and the temperature of the air inside the chamber. When I remembered the lab study of Kristina Arnebrant, the young researcher from Sweden who showed that alder delivered nitrogen to pine through mycorrhizal connections, I returned the next day to collect foliar samples from the birch and fir to be tested for nitrogen concentrations.

A couple of weeks later, this data came back from the lab. The birch had double the concentration of nitrogen in its leaves compared to the fir needles. Not only did this help explain the higher photosynthetic rates of birch compared to fir (nitrogen is a key component of chlorophyll), it also meant there was a nitrogen source-sink gradient between the two species. Like between the nitrogen-fixing alder and nonfixing pine in Kristina's study.

I wondered if this nitrogen source-sink gradient might be as important as the carbon source-sink gradient in driving the flow of carbon from birch to fir. Or if the source-sink gradients in the two elements worked hand in hand. Instead of carbon flowing in whole sugar molecules through the fungal pipeline, the sugars might break down into their bare elements (carbon, hydrogen, and water) and the free carbon could join up with nitrogen vacuumed from the soil to form amino acids (simple organic compounds ultimately used to make proteins), for example, in leaves and seeds. The newly formed amino acids, and any lingering sugars, would then shoot through the network. With gradients in both carbon and nitrogen—carbon in sugars and nitrogen plus carbon in amino acids—birch was perfectly equipped to shuttle more food to fir than it received in return.

The month I needed to wait for fir to slow down in the shade of the tents seemed to crawl by. I went hiking with Jean along the Stein River, visiting the Asking Rock and wriggling our toes in the glacial water. I spent days with my crew measuring trees in our other experiments; I checked my messages to see if Kelly had called. Dad said he and Tiffany were fine, but I still wanted to hear from him. As the days ticked by, I imagined the gradient in photosynthetic rate between birch and fir growing steeper. One week, two weeks, three weeks crept by. The physiology of the firs in deep shade, I thought, must be as slow by now as flies in the cold. When my four-week furlough ended in the middle of July, it was time to find out if the paper birch and Douglas fir were communicating.

I returned to the site with my university research associate, Dr. Dan Durall, an expert at labeling trees with carbon isotopes. He was also my next-door neighbor in Corvallis. Dan had just finished a project for the Environmental Protection Agency where he'd labeled trees with carbon-14 and learned that half of the carbon was shuttled and stored belowground—in roots, soils, and microbes such as mycorrhizal fungi. The EPA needed this information so they could start figuring how best to store carbon in forests for mitigating climate change. It was the early 1990s, and I'd heard about climate change in a noon-hour seminar at Oregon State University and was thunderstruck to learn of the catastrophe being predicted. When I'd returned to Canada with the news, the managers at the Forest Service hadn't believed me.

Our first job was to set up a tent at the site because the mosquitoes were as big as snipes. The air was so dense with them—and blackflies, deerflies, horseflies, and no-see-ums—that each breath brought in a fluttering insect. We hauled in a table-turned-lab-bench for assembling our equipment and handling our samples. In the time it took me to race to the truck to grab the syringes and gas tanks and run back to the tent and zip up the door, my face was raw with bites. I was glad when the tent was up and the equipment organized. Without the shelter, we'd be pulp from the insects. With it, we'd survive—barely.

Labeling the seedlings would take us six days—ten triplets per day. In each triplet, we'd place one garbage-sized clear plastic bag over the birch and one over the fir. In half the triplets, we would inject the bag over the birch with carbon dioxide tagged with the carbon-14 isotope, and we'd inject the bag over the fir with carbon-13-tagged carbon dioxide, which they'd absorb over a couple of hours through photosynthesis. This would allow us to detect carbon moving in both directions between the trees. Carbon-13 and carbon-14 are slightly heavier forms of the common element carbon-12—with atomic weights of thirteen and fourteen instead of twelve—but they are naturally very rare and can therefore be used as tracers of how carbon-12 behaves in photosynthesis and sugar transport. In the other half of the triplets, I'd switch which tree got what—tagging birch with carbon-13 and fir with carbon-14—in case the different isotopes were distinguishable by birch and fir and therefore affected how much they took up through photosynthesis and how much they transmitted to their neighbor. If trees did detect the tiny difference in the mass between the two isotopes, I could calculate the relative magnitude of transfer of each isotope and then correct for their subtle discrimination differences to be sure it didn't mess with my ability to detect how shading affected the carbon fluxes.

Dan and I talked about making sure that the carbon isotope that Douglas fir picked up from birch wasn't tagged carbon dioxide that had escaped into the breeze when we removed the bags after the two-hour labeling period. I was so focused on what moved through the mycorrhizal networks that I really didn't care that much about minuscule amounts that might drift through the air. Besides, cedar was my control and it would pick up a mix of aerial and soil transfers of carbon, and it would tell me in total about any errant escapees.

But Dan insisted we could do better than that. Before removing the bags, we could vacuum out the unabsorbed isotopic carbon dioxide and capture it in tubes. Then we would largely eliminate potential aerial transfer.

After so much planning, I was in a fever to get started labeling my seedlings. This was the most daring experiment I had ever done, with so much potential to change how we view forests, and yet at the same time it held the possibility of turning up nothing. It felt as if I were about to parachute out of a plane, maybe land on Easter Island. I was jittery with adrenaline. Once I had my results, even if we were still out of touch, I'd show Kelly the prize in person. I'd go visit him and Tiffany, our bar fight be damned.

The next day, inside the tent, we tested the method we'd devised for labeling the seedlings with carbon-13. I'd purchased 99 percent pure ^{13}C-CO_2 gas straight from a special supplier, and it had arrived by mail in two gas cylinders the size of corncobs. Each cylinder of gas cost one thousand dollars, consuming 20 percent of my budget. To practice extracting the ^{13}C-CO_2 gas from the cylinders, Dan took one and screwed a regulator onto it, then clamped a meter-long piece of latex tubing onto the outlet spigot. The idea was to release the gas slowly into the tubing as though we were blowing up a sausage-shaped balloon. Once the tube was full of gas, we'd use a large syringe to withdraw fifty milliliters of the ^{13}C-CO_2 gas, which we'd then inject into the plastic bag over the seedling so it could absorb the gas through photosynthesis, and perhaps transmit some of the isotope through the mycorrhizal fungi to its neighbor. My job was to make sure the clamp on the end of the tubing held tight while he turned the tap on the cylinder to fill the tubing with gas.

"Ready?" he asked, sweat dripping off his eyebrows as he hovered over the lab bench.

"Ready," I replied. I nervously tightened the clamp. I'd done okay in my chemistry labs at the university, but being in the bush with these chemicals terrified me.

Dan turned the knob on the regulator.

"What's that hissing noise?" I asked. The tubing was on the ground, wriggling like a snake with the thousand dollars' worth of gas spurting out the end. My clamp had come undone under the pressure. I tied a knot in the tubing as the last bit of gas hissed out.

Dan's mouth hung open. I looked at him as if I'd just dropped a Ming vase.

I was glad we had two cylinders.

We perfected our techniques for injecting the isotopic gases into the bags, and the day came when we were ready to label the seedlings. It was warm out in the clear-cut, and even hotter inside my plastic suit. Since carbon-14 is radioactive, I was worried about exposure, so I wore a rainsuit, a respirator, giant plastic goggles, and rubber gloves sealed to my sleeves with duct tape. Dan thought I was nuts as he donned his simple white lab coat, knowing that carbon-14 was not very dangerous in the way we were using it. The energy of the particles was so low they'd barely penetrate a single layer of skin. Easily stopped by a pair of surgical gloves. The scariest thing about carbon-14 is that if it does manage to stick to you, maybe lodge in your lung, it hangs around for a long time, given its half-life of 5,730 (+/- 40) years. Carbon-13, on the other hand, was a nonradioactive isotope and of no concern.

At the first set of triplets, I took the tent off the Douglas-fir seedling and placed a tomato cage over it and a second one over the paper-birch seedling, leaving the cedar seedling in the open. The cages were to serve as frames for the plastic labeling bags, ensuring that the sides of the bags stayed inflated during the labeling period.

With the cages up, we were ready for the moment I'd been planning and waiting for during the past full year, ever since I'd first put the seedlings into the ground. To see if birch and fir traded carbon—whether they communicated with each other through belowground networks. It felt like a defining juncture—whether my intuition was right that collaboration in the forest was important to its vitality. And if so, I had a big responsibility to stop the madness of the wholesale removal of native plants. Over the top of the first set of tomato cages, we draped the gas-tight bags—as though placing a curtain over a parrot cage—completely covering the individual birch and fir. We sealed the bottom of the bags around the stems of the seedlings and legs of the tomato cages with duct tape, making sure there were no leaks. Immediately before the last strip of duct tape was battened down, Dan reached inside one of the bags and taped a frozen vial of radioactive sodium bicarbonate. He carefully injected lactic acid into the frozen radioactive solution using a large glass syringe inserted into a porthole in the bag. Once he plunged the needle into the portal, the acid slowly dripped into the frozen vial, releasing ^{14}C-CO_2 for the paper birch seedling to absorb through photosynthesis.

In the meantime, I was back at the tent, extracting fifty milliliters of ^{13}C-CO_2 from the corncob cylinder into a syringe so I could inject it into the other bag covering the Douglas fir. Sweating so profusely my goggles kept fogging, I waddled from triplet to triplet, making my injections while Dan made his. Mosquitoes and flies swarmed like dust. Dan moved swiftly between the triplets and our lab bench, where we kept the vials of radioactivity frozen in liquid nitrogen. I lagged behind as I extracted each syringe of ^{13}C-CO_2 at the lab bench, then shuffled to the next triplicate.

After leaving the seedlings to absorb the labeled carbon dioxide gas for a couple of hours, we vacuumed out any potential excess iso-tope, and then we pulled off the bags. Whatever traces of gases that might have remained were quickly carried by the gentle breeze into the atmosphere.

The bags off, Dan ran back to the lab tent to escape the clouds of bugs. I waddled as fast as I could behind him, zipped the door closed, and tore off my plastic armor. Like a surgeon, Dan peeled off his latex gloves and dropped them into the garbage bag of used gear. We stared at each other. "We did it!" I exclaimed.

Dan said, "Maybe." We still needed to check the seedlings with the Geiger counter.

Right. I pulled my plastic suit and surgical gloves back on, grabbed the Geiger counter, and raced back to the closest triplet. The breeze had picked up, and the leaves of the birch seedlings were whiffling around their twirling petioles, the fir leaning steadily into the air cur-rent. Across the lake, storm clouds were piling up like inky cap mush-rooms. A squirrel ran in front of me and stopped on a stump to watch.

I lifted the Geiger counter up to the leaves of the carbon-14-labeled birch. I held my breath—were they radioactive? If not, then all of our preparations would be for naught. If the donors hadn't absorbed the radioactive carbon dioxide, then we wouldn't know if they could then transmit organic compounds to neighboring firs. Dan appeared next to me, looking anxious.

I flicked on the switch. A crackle sang from the wand. Dan's face lit up. The hand on the meter swung hard to the right, showing a high radiation count.

"Oh, good. I did it right," Dan said, relieved.

"Do you think we'll be able to detect anything in the fir neighbor?" I asked.

"I doubt it. It's only been a few hours since we started the labeling," he said, trained to be cautious with early results. Based on Read's study, it would likely take a few days for the radioactivity to transmit belowground from the birch to the fir. Even if it did move into a neighbor, the amount would probably be below the detection limit of the Geiger counter, and we'd have to wait for results until we ran the samples at the lab.

But what would it hurt to check with the Geiger counter now? We could try to get a reading in advance to see if there might be a hint that the Douglas-fir needles held the answers. I calmed myself down. I was sure Dan was right; he knew more about labeling plants than most anyone.

But what the heck. It cost nothing to try. I instinctively walked to the neighboring fir and knelt. Dan couldn't help but follow me and lean over my shoulder. We both inhaled the piercing scent of fir-needle resin, and for a moment I forgot about the years of difficult work and bouts of frustration. I brushed my hand over the end of the wand, to make sure nothing was there to mask the signal. The moment of truth had come. The conductor raised her hands to the orchestra, the ensemble readied their instruments. Tilting my ear toward its stem, I ran the Geiger counter over the fir's needles.

My wrist lifted slightly in an upbeat, and my Geiger counter wand crackled faintly as the dial on the meter swung up a tad. Strings and woodwinds, brass and percussion, exploding as one, flooding my ears, the movement allegro and intense, concordant and magical. I was enraptured, focused, immersed, and the breeze sifting through the crowns of my little birches and firs and cedars seemed to lift me clear up. I was part of something much greater than myself. I shot a look at Dan, his mouth frozen open.

"Dan!" I cried. "Did you hear that?"

He stared at the Geiger counter. He had wanted with his whole heart for the labeling to work, and what we were hearing from the fir was beyond anything he'd expected.

We were listening to birch communicate with fir.

C'est très beau!

We wouldn't know for sure until we properly analyzed the tissue samples with a scintillation counter, more sensitive at detecting radioactive carbon-14, and a mass spectrometer, for measuring carbon-13. These would quantify accurately how much labeled photosynthate had traveled between birch and fir. Still, Dan's eyes were ablaze at this initial clue. Me, I was over the moon, ecstatic, the grin on my face irrepressible. I threw my arms up to the wind and I shouted, *Yes!* Deep down, in our own ways, we both knew that we'd picked up something miraculous happening between the two tree species. Something otherworldly. Like intercepting a covert conversation over the airwaves that could change the course of history.

I stepped to the cedar in the triplet, my palms sweating. I seemed to own the answer already. I lifted the wand and ran the Geiger counter over its braids.

Silence. Cedar was in its own arbuscular world. Perfect.

How long the isotopes would take to move fully, to complete their journey from seedling to seedling, was a mystery, so I planned to wait six days. Enough time for more of the isotope to wend from the roots of the donor through the fungus to arrive in the tissues of the neighboring seedlings. I sank down, and Dan sat beside me, our instruments in our laps, the movement of the breeze slowing, a lone meadowlark singing. My feelings of frustration and rejection over my work, my struggles with grief and self-loathing over quarreling with Kelly evaporated for that moment. I put my arm around Dan's shoulders and whispered, "We've found something really cool here."

AFTER THE SIX-DAY WAIT, we dug the trees out of the ground. The roots of the birch, fir, and cedar were massive, intertwining, and covered with mycorrhizas. "Looks like a herd of gophers has been here," I said after we finished harvesting. We separated the roots and shoots into separate bags and packed up the mosquito tent and table-turned-lab-bench.

As we drove away, I looked back at the tiny piece of land that was going to tell us just how much our seedlings were connecting and communicating with each other. A raven flew over and called in a low croak. I remembered that the Nlaka'pamux, on whose land we had performed this experiment, see the raven as a symbol of change.

The next day, I drove to Victoria with the samples in coolers. I was using a designated lab facility to grind my tissue samples into powders, which I would send to a lab at the University of California, Davis, for analysis to detect the amount of carbon-14 and carbon-13 in each tissue sample. I ground my radioactive samples in the fume hood, a special enclosed cupboard with a glass window and an overhead duct to draw out the air, so that any radioactive particles would be sucked out of the cabinet and expelled safely to some hidden chamber for proper collection and disposal. Grinding the tissues was tedious and awkward; I had to put the grinder, a metal machine the size of a coffeepot, inside the fume hood to suck up the wood dust and avoid spreading the radioactivity through the lab. And to prevent me from being covered with the dust or breathing it.

The first day, I checked into the lab at eight a.m., donned my lab coat, put on my safety goggles, strapped down my dust mask, filled the grinder with a root sample, and leaned into the fume hood. Hour after hour, I ground the samples as finely as I could. At five p.m., I stacked the samples I'd ground that day in a box. I cleaned the fume hood, lab bench, and floors, ran the Geiger counter over the surfaces to make sure no radioactive particles lingered, washed myself, and checked out of the building. I went to my hotel room, showered, ate a hamburger at the pub next door, flopped onto my bed, and fell asleep with the television on. For the next four days, I'd wake to my six a.m. alarm and do it all again.

It took five ten-hour days to grind all the samples. On the last day, I was vacuuming out the fume hood and fiddling with my dust mask when I noticed the metal tab on the top covering my nose. When I squeezed the sides, it miraculously tightened the mask over my nose. My heart sank. I hadn't been squeezing the nosepieces on the dust masks properly.

I ripped off the mask and stared at a thin film of dust inside. I pulled a film of wood dust out of my nose and almost fainted. I had been breathing in the tiny ground particles. I dropped onto the stool in the lab in disbelief.

There was no way to undo my mistake. What was done was done.

I called Dan, and he reassured me that it was unlikely I'd breathed any into my lungs, and if I just washed myself well I'd be fine. I hoped he was right. I went to the eyewash station and flushed my eyes, nose, and mouth. I put away the last of my equipment and packed the rest of the samples in boxes to be shipped to California.

A FEW MONTHS LATER, I was at Oregon State University, crunching the isotope data that had come back from the California lab. My tiny, windowless office was a former-insect-rearing-laboratory-turned-hideaway. The overhead heat lamps were long disconnected, and gas spigots stuck lifeless out of white tile walls. Don was working on his own dissertation, examining clear-cutting effects on forest composition and carbon-storage patterns across a portion of British Columbia the size of Oregon, soon to discover that clear-cutting was causing carbon dioxide to pulse into the atmosphere at unprecedented rates. Our world had collapsed into data analysis, running, and drinking beer with other grad students.

When I wasn't analyzing the isotope data, I was in the microscopy lab, examining the root tips of Douglas-fir and paper-birch seedlings for mycorrhizas. I'd grown paper-birch and Douglas-fir seedlings in a separate greenhouse experiment using soil I'd collected from my field site. Some of the birches and firs grew isolated in separate pots of soil, and others together in the same pot. After eight months of watering and watching, I harvested the isolated singles and examined their root tips under the microscope. The spores and hyphae in the soil had colonized some of the root tips. Even though the birch and fir had been grown separately, most of the mycorrhizal fungi colonizing their roots were *the same ones*. Not just one species of fungus, but *five*. Fungi as varied as the mushrooms they spawned.

Phialocephala, with its eerily translucent dark hyphae running inside and outside the roots of both the birch and fir.

Cenococcum, with its jet-black mantle coating a smattering of the root tips and emanating bristles as stout as a hedgehog's.

Wilcoxina, with its smooth brown mantle and see-through mycelia emerging from delicate beige mushroom caps.

Thelephora terrestris, forming white creamy root tips and blooming into fanning rosettes of tough brown flesh with white margins.

Tiny but prolific *Laccaria laccata,* with its bland root tips and snow-white emanating hyphae that merged into bald orange-brown mushroom caps.

When I got to the birch and fir grown in pairs, my face was warm with anticipation. Earlier studies had shown that different species grown in groups spawned whole new mycorrhizal species that could not form on either tree grown alone. It was as if they needed to prime each other, egg each other on, maybe by providing the neighbor with carbon transmitted through fungal linkages.

When I put the fir roots from the mixed pots under the microscope, I almost fell off my lab stool. The roots looked as big and abundant as the strands on a kitchen mop. More strikingly, the different fungal species colonizing them were as diverse as tree species in a tropical forest. Not only that, two brand-new species emerged on the fir *and* the birch: *Lactarius,* with its white creamy mantle, the same color as the milky fluid dripping from the gills of its milky-cap mushrooms. And *Tuber,* covering the root tips with blond chubby fungal clubs and spawning black underground truffles, similar to a Périgord truffle.

I ran to the office of my doctoral supervisor, Dave Perry, his head bowed at his computer, and when he looked up, he pushed his reading glasses back on his head of long gray hair. Not a square inch was visible on his desk, owing to the teetering stacks of papers he'd accumulated over the decades. I shouted that the firs grown with birch looked like decorated Christmas trees. While the firs grown all alone had fewer mycorrhizas.

"Oh, man," Dave said as he jumped up and gave me a high five. He nodded along as I described the colorful fungi in the mixed pots, gesturing in excitement at how big the roots were. He'd already seen Douglas fir share fungal species with ponderosa pine, but he didn't know if they connected the trees or transmitted nutrients. Dave and I both knew these results meant birch and fir had the potential to form a robust, complex, interlinking network. But more importantly, as I suspected from analyzing the isotope data from my field experiment, we knew we were on the cusp of discovery whether the trees communicated through the network. Dave pulled a bottle of Scotch whisky from his desk and poured an ounce into each of two beakers. He loved seeing his students make their first eye-popping discovery. I imagined birch and fir weaving a network as brilliant as a Persian rug.

The seven shared fungi, we'd later discover, represented a fraction of the dozens of fungal species in common between birch and fir. The cedars, as I expected, were colonized only by arbuscular mycorrhizal fungi and were not part of the network joining paper birch and Douglas fir.

WHEN THE FIELD carbon-transfer data came in from the lab, I held my breath. This was it. The science was sound. The experiment had taken every variable into account. I was alone in my windowless office as I scanned the report. My cheeks burned as my eyes raced up and down the data columns. I ran the statistical code to compare how much carbon-13 and carbon-14 had been absorbed by the birch and fir, and whether shading fir made a difference in the amount. Again and again I checked the numbers, just to make sure. I sat in disbelief. Paper birch and Douglas fir were *trading* photosynthetic carbon back and forth through the network. Even more stunning, Douglas fir received far more carbon from paper birch than it donated in return.

Far from birch being the "demon weed," it was generously giving fir resources.

The amount was staggering—it was large enough for fir to make seeds and reproduce. But what really floored me was the shading effect: the more shade that birch cast, the more carbon it donated to fir. *Birch was cooperating in lockstep with fir.*

I re-analyzed the data over and over to make sure I hadn't made a mistake.

But there it was, telling me the same thing, no matter how I looked at it. Birch and fir were trading carbon. They were communicating. Birch was detecting and staying attuned to the needs of fir. Not only that, I'd discovered that fir gave some carbon back to birch too. As though reciprocity was part of their everyday relationship.

The trees were connected, cooperating.

I was so shaken I leaned against the tile walls of my office to absorb what was unfolding, because the earth seemed to be rumbling. The sharing of energy and resources meant they were working together like a system. An intelligent system, perceptive and responsive.

Breathe. Think. Absorb. Process. I wanted to call Kelly, but we were still at our impasse. We'd be back in touch soon enough.

Roots didn't thrive when they grew alone. The trees needed one another.

I sorted through a stack of papers documenting the competitive effects of trees on one another, next to a growing pile of papers on how trees facilitated one another, which I collected because it frustrated me that researchers were firmly split into camps. Fights erupted in seminars. Each held some piece of the truth, but still to emerge was the full complexity of the interactions among trees. Regardless of the disagreements, the indiscriminate removal of native plants was continuing, and the diversity of the forest was still falling victim. I had a choice: I could show policymakers all of this, taking the chance that they would try to suppress me. Or I could stay in my lab, hoping someone would eventually use my findings.

The office phone rang.

I left my desk to answer it, though the phone here was almost never for me.

I picked up.

As if from a distance, I heard Tiffany sobbing before she blurted, "Suzie, listen. Kelly is dead."

I clutched the edge of the desk, my ear pressed against the receiver.

Tiffany's words rolled out in fits and spurts: Changing the sprinkler heads. Returning his tractor to right outside the barn—shoved it into park—left it idling—ducked under the barn door—barn door crashed in—crushed him against a dump truck.

I listened, frozen.

She told me Kelly had a premonition. Just last Friday, he had been herding cows from the alpine to lower pastures. The grass was frozen, streams iced over, and the cattle huddled in the November fog. He peered into the mist at a cowboy drifting toward him. He had been grateful to see him; it was a big job to herd fifty head with his horse and Nipper, his border collie.

He looked again; it was an old friend who gently tipped his worn hat his way and smiled under his gray mustache. He rode easily in his saddle, chaps warming his long legs.

Suddenly Kelly shuddered.

He knew the cowboy. But he had died the previous year.

The old man beckoned, and Kelly followed. The dead rancher rode slowly through shifting fog. Kelly, in disbelief, spurred his horse to catch up. The cowboy turned his head and looked back to make sure Kelly was coming. He was.

As quickly as he had appeared, the old man slipped into the mist.

Kelly had been terrified. Tiffany began to wail, "I sat with him in the hospital, his body was cold. How can he leave me?" Their baby was due in three months.

After her call, I couldn't hear anything, as though all sound had stopped. Time collapsed. I couldn't stop shaking. Don was out playing baseball, but I didn't know where. I made my way home, stunned. I had to make the calls—my mother, father, sister, grandparents—but waited until Don came home, and he helped me let everyone know, reliving the shock with each one. Like being punched in the face over and over.

The next day, I flew back to Kamloops. My senses were numbed, as though I was in an old silent movie.

The funeral was held in agonizing cold. The aspens were bare, the firs nestled beneath their dendritic crowns drooping in snow. Tiffany held her arms around the son growing in her belly, her skin like porcelain, her face serene in her grief. I wanted to stand near her, just to be with her, but I was busy with Mum and Dad. Robyn was also six months pregnant, and she stood with Bill and Tiffany at the back of the church. Kelly's friends gathered, cowboy hats shading their eyes, telling stories about what a good man he was, about their times together. The pews, here long before any of us were born, to be there long after we'd all gone, held in their wood a solidity none of us could match, a solemnity we could only try to absorb. Kelly lay cold in a simple pine box. I couldn't breathe. I wanted to kiss his forehead but I couldn't bend. I was sick with remorse. I could never make amends. We would never reconcile. Our final words brutal parting shots in drunken anger and misunderstanding.

Brother and sister. Shattered forever.

9

QUID PRO QUO

My grief came in waves. Tears, regret, anger. Don was still in the United States finishing his dissertation, so I was alone. Our Corvallis neighbor, Mary, called to comfort me, to tell me this kind of pain takes time, and I appreciated her kindness. But my sorrow was unrelenting. I couldn't focus on work, so I cross-country skied. Day. Night. Long, punishing ski tours. Beating myself up, but doing so out in the forest, which I knew on some level, even in my anguish, held the promise of healing.

Sometimes, when the worst happens, we are no longer afraid of the things that used to scare us. The small things. The things that aren't a matter of life and death. I threw myself into my research, if only to bury my despair at what I couldn't repair, trying to find, in my connections to the trees, what I'd lost forever with my brother. I'm not sure if it was because of Kelly or in spite of him, but I decided to publish my research findings. With the encouragement of Dave and Dan and the rest of my doctoral committee, I sent an article to the journal *Nature*.

A week later, I received a letter from the editor. He had rejected it.

The criticisms seemed simple enough to correct, and I had nothing to lose, so I revised the article and resubmitted it, the way I used to throw driftwood back into Mabel Lake when it kept returning to shore. The way Kelly and I used to keep fixing our homemade raft so we could explore the creek the next bay over.

Nature decided to publish the new version as the cover story in

August 1997. They used one of Jean's photos of a mature mixed birch-fir forest near Blue River. I was astonished. My article had beat out the discovery of the genome of the fruit fly for the front cover. The journal also invited Sir David Read to write an independent review of my article, which they published alongside mine in the same issue. He wrote: "The study of Simard *et al.* . . . [addresses] these complex questions in a field situation and for the first time . . . shows unequivocally that considerable amounts of carbon—the energy currency of all ecosystems—can flow through the hyphae of shared fungal symbionts from tree to tree, indeed, from species to species, in a temperate forest. Because forests cover much of the land surface in the Northern Hemisphere, where they provide the main sink for atmospheric CO_2, an understanding of these aspects of their carbon economy is essential."

Nature called my discovery *the wood-wide web,* and the floodgates opened. Calls from the press kept my phone ringing off the hook and emails cramming my in-box. I was as stunned as my colleagues at the attention the *Nature* publication brought. One night, my own floodgates burst, and I sobbed . . . not something we often did in my family. I had hidden my grief to let my parents express their own in peace, but now I couldn't stop it from flowing, crying until I couldn't anymore. I pulled myself together when *The Times* in London called me, then *The Halifax Herald*. I got a note from France, a letter wrinkled and postmarked from China.

Under the global spotlight, the Forest Service might notice.

I couldn't save Kelly. But maybe I could save something.

ONE AFTERNOON, Alan leaned against my office doorjamb. The winter had dragged on, and I was low. Although it had received international press, the *Nature* publication hadn't budged the Forest Service's policies, leaving me uncertain where to focus my efforts next. Alan told me to put on my boots and get back in the field to sort myself out. When I felt better, he said, we'd take the policy guys out to the forest to show them what this research meant. I grabbed my keys and headed to my mixtures experiment, where the old rancher had tried to thwart me by seeding in grasses.

I turned my pickup off the Trans-Canada Highway at Eagle River.

Untracked slush coated the gravel road, and I knew I was the first person to have come along since the fall. I arrived at the old birch where I'd collected soil to revive my seedlings. I got out my flip phone to call Robyn, but there was no service. She was due in a few weeks, as was Tiffany. I wanted my own children, but Don was still completing his dissertation in Corvallis, due to finish at a time that a memorial was planned for Kelly at the beginning of the Williams Lake Stampede; just as well, since Don mixing with cowboys was like blending oil with water.

I gave up on the phone and grabbed my cruiser vest and bear spray, and walked the last kilometer in. I needed sharp, wet air in my lungs, to feel *something* real. To walk among the trees, smell the flowing sap, feel their presence, let them know I was here, listening.

At the old-growth forest next to the mixtures experiment, I waded through foot-deep snow, rain pants heavy. A wisp of cloud, a faint glow, beckoned. Strands of creamy lichen hung from branches, like Kelly's white shirts still hanging in Tiffany's closet. Deep inside these woods, I'd established the second of my doctoral field experiments. I'd planted twenty clusters of five Douglas firs each under the dense forest canopy to see how the seedlings might survive in the deep shade, how long they might live in the gloom. In half the clusters, the roots of the five little seedlings were free to interlace with the mycorrhizal network of the ancient trees. In the other ten clusters, I'd blocked the roots from the elders by encircling the fledgling group with a meter-deep band of sheet metal. Just as I'd done with the triplets of fir, birch, and cedar for my wood-wide web experiment, though here in the dark shadow of the canopy, I'd planted only fir. Inside the timberline, the possibility of fir seedlings connecting and communicating with their old neighbors was even greater.

Where new growths, clustered near parents, clung to survival.

Where linking into the mycorrhizal network of century-old trees might mean the difference between life and death.

Where the elders could sustain the young so they'd be poised to fill the tree-fall gaps when the old passed on. Imparting a head start to new generations. Given how shady this understory was, I guessed that the giant trees would be a much stronger source of carbon for the tiny firs here than the birch seedlings had been to the firs of the triplets I'd labeled in the clear-cut. Niagara Falls compared to a babbling creek. The source-sink gradient here was extraordinarily steep, in synchrony with the roles of these old-growth sentinels.

At the first cluster of firs inside the thick forest, there was only one survivor, its sickly yellow leader barely sticking above the cover of snow. I'd loved this experiment, but it seemed doomed. My throat felt like stone; my heart ached. Icy water dripped from the canopy and ran in rivulets down my neck. Snow-lined drooping cedar boughs reminded me of bleached fish skeletons. The faint luminescence of skunk cabbages awakening in the fens of humus barely broke the pallor.

Shivering, I pawed the snow off the survivor, so young but near the end of its time. I brushed ice crystals off other blackened stems, four of them, their dead roots imprisoned. I felt around and found the circle of sheet metal I'd wrapped around the group to estrange them from the old trees, to test what I suspected—that I was making a grave. It was here in the dark understory that connection to family seemed most crucial.

I walked through the mist to the next cluster, checking my hand-drawn map. A group of green leaders emerged from the snow. I'd planted these seedlings without a barrier, leaving them to connect with the rich fungal network of the elders. All had put on a centimeter of new growth since last summer, and each had a fat new terminal bud. I scraped away the snow, shallow here because of the warming stems, and peeled back the centimeters-deep litter. Thick, richly colored mycorrhizas like a Renaissance painting wended through the organic horizon, and I suddenly felt lighter, hopeful. I uncovered the root of a seedling and traced a dark *Rhizopogon* strand connecting it to a giant Douglas fir a few meters away. Another root was coated in a shimmering yellow mycorrhizal fungus, a *Piloderma,* and I followed the fleshy yellow threads to an old birch. I sat back, startled. This little seedling was entwined in a prosperous mycorrhizal network with both the mature Douglas fir *and* the paper birch.

I tugged my hat over my ears. The network did indeed seem to be sustaining the seedling. The old trees could be sending it sugars or amino acids through the fleshy fungal mats, to compensate for the minuscule rates of photosynthesis the tiny needles could muster in the dim light and the sips of nutrients the fledgling roots were pulling from the soil. Or maybe the old trees just inoculated the seedlings with their own diverse suite of mycorrhizal fungi so the youngsters could get at the tightly bound soil nutrients without additional assistance.

I dug through the forest floor encasing another of the seedlings and found a half-dozen more mycorrhizas on its roots. By now, I knew that there were more than a hundred species of mycorrhizal fungi in this forest. About half were generalists, colonizing both paper birch and Douglas fir in a diverse network. An intricately woven rug. The other half were specialists, with fidelity to either birch or fir, but not both. Each of these specialists was thought to have its own niche. Some were good at acquiring phosphorus from humus, others nitrogen from aging wood. Some sopped up water from deep in the soil, others from shallow layers. Some were active in spring, others in fall. There were some that produced energy-rich exudates that fueled bacteria performing other jobs, such as breaking down humus or transform-ing nitrogen or fighting disease, while other fungi produced fewer exudates because their jobs required less energy. The glossy sheen of the *Piloderma* mycorrhiza I'd seen linked to the birch suggested that it held a rich supply of carbon, supporting a biofilm of the luminescent *Pseudomonas fluorescens* bacteria, whose antibodies could shrivel the growth of the root pathogen *Armillaria ostoyae.* The *Tuber* mycor-rhiza turned out to host *Bacillus* bacteria that transformed nitrogen, helping explain why birch leaves had so much more nitrogen than fir needles.

But we knew almost nothing of the functions the vast majority of mycorrhizal fungi performed. What was known was that older forests had a greater variety of fungi than plantations did, and these suites of species especially associated with old trees were thick and fleshy and robust and able to access resources sequestered in hard-to-get-to corners of the soil. They unlocked essential nutrients that had been held tightly for centuries, in tenacious complexes of humus and min-eral particles. Atoms of ancient nitrogen and phosphorus that were sequestered on phyllosilicate clays and bound in carbon rings linked together like chicken wire.

By collecting mushrooms over the seasons and years, Dan and I had figured out that the old-growth forests contained special old-growth fungi. Some only showed up in particularly rainy months and years, and some appeared only once. Others fruited only in dry months, while some mushrooms flushed no matter the season. We'd also dug up roots of birch and fir in forests ranging from a few years to hundreds of years old. We'd analyzed their DNA and compared it to data in a universal genetic library to determine the fungal species.

I walked deeper into the forest, where hemlocks and spruces mingled under the firs and birches, and stopped at a sapling shedding its parka of snow. After I swept the last crust of melded crystals away, its supple stem slowly straightened. *We are built for recovery,* I thought. I stopped at some hemlock saplings marching in single file along a nurse log, just as I'd seen at Mabel Lake. I figured this gave them all manner of advantages—an escape from soil pathogens, or a ladder up to the light. Roots of the hemlock saplings grew over and under the crumbling logs, and they enwrapped the gnarls of tree roots and sprawling rhizomes of hazelnuts and Sitka mountain ash and false-box with the intimacy of a tightly knit town. They were probably all connected in a shared network of ectomycorrhizas. Even the western red cedar and yews, and the ferns and trillium, which by now I knew were arbuscular mycorrhizal, probably formed a network. A seamless arbuscular mycorrhizal web entirely separate from the ectomycorrhizal one. Regardless of the presence of discrete mycorrhizal networks, all the plants in this forest belonged to one another.

I now knew that birch and fir were connected and communicated, but it didn't make sense that birch always gave more carbon to fir than it received in return. If this were always so, fir might eventually drain the life out of birch.

Were there times in its life when fir might give to birch more than it received? Perhaps when the forest was older and fir had naturally outgrown birch there was a net transfer of carbon from fir to birch.

Slatted light guided me to the timberline with the neighboring clear-cut. My third doctoral field experiment was here, where the rancher had spread the grass seed to carry out his revenge. I was lucky the trees had grown well in this little area in spite of the grass. The saplings, now five years old, were already taller than I was. I crouched at one birch sapling surrounded by a thick lip of plastic poking from the ground, part of the meter-deep wall I'd sunk down to encircle its root system. This setup was similar to the way I'd used sheet metal in the forest. Instead of building a moat around groups of seedlings, though, I'd built separate ones around each of sixty-four seedlings planted in a grid, a small forest. The plastic was still strong and intact, and it would maintain its integrity for many years. I was testing whether birch continued to help fir through the childhood years, and whether fir eventually gave back—perhaps in the off-seasons of early spring and late fall, when birch had no leaves—and did so even more as fir slowly, naturally overtook birch in early adulthood.

To figure this out, I was comparing this trenched plot of trees with a nearby plot of sixty-four birch and fir left untouched, interwoven as one. Making the trenches was like conducting an archaeological dig in an ancient city of stumps. Barb and I had hired a fellow with a mini-excavator and a crew of four young women with shovels to dig the meter-deep trenches. We'd clawed out sprawling root systems and edged out granite boulders to make nine trenches along the eight rows of trees, the ninth trench being on the outside of the last row. We'd excavated nine more in perpendicular rows, resulting in a crisscross. Popping out of the maze of trenches were the sixty-four islands of soil containing one tree each. When we'd lined the islands with plastic—so the roots and mycorrhizas wouldn't burst through—and backfilled the maze with dirt, the only things left visible were the slivers of plastic breaking the surface. Hidden below was a perfect eight-by-eight Latin square.

I wondered if the firs really were smaller here than in the other plot where the roots were free to mingle with those of the neighbors. One sapling was dead, its red needles in the snow like drops of old blood, so I grabbed its flaking trunk and pulled it out of the ground. The rotting root stubs were covered with creeping black fungal strings—rhizomorphs. Flicking open my knife, I carved off the bark at the base of the stem and laid bare the woody xylem. Snow-white fungal hyphae formed a noose, confirming death by the fungal pathogen *Armillaria ostoyae*. I searched the plastic trenches for more corpses—a third of the firs had died.

In the untrenched plot, all were alive, and I could have sworn bigger too. The wings of a raven swooshed, and a train's whistle pierced the air. I pulled out my calipers and notebook and measured the diameters of all the birch and fir trees in both plots. When the sun slid behind the mountain, I made my way, soaked and shivering, back to my truck. I turned the ignition, twisted the heat up full blast, and crunched the data with my calculator in the waning light.

My guess was right. The firs connected with their birch neighbors were not only all living, they were larger than the trenched firs. The birches, on the other hand, were unaffected by their intimacy with the firs, not drained by the association. Birch was not being bled dry by transmitting some of its carbon; it was giving enough to boost the survival and growth of the fir without cost to its own vigor.

Could birch shut off the tap when it sensed the fir was no longer in need? And my question persisted about whether birch also benefited from fir, perhaps at some other time or in some other way, in a manner not yet evident from these simple measurements. None of the firs showed any signs of *Armillaria* root disease. Growing enmeshed with birch seemed to protect fir from sickness—as I'd seen in many other experiments. I'd convinced Rhonda, my summer field assistant at the Forest Service, to do her master's continuing my investigation of *Pseudomonas fluorescens,* the fluorescing bacteria I'd found antagonistic to *Armillaria ostoyae*. She'd compared the abundance of the helpful bacteria between forest types and found four times more in birch stands than fir stands, probably because birch roots and mycorrhiza fungi, fueled by higher rates of photosynthesis, provide more food for the bacteria than firs can manage. She also found as many bacteria on firs as birches when the two species were mixed together, as though the tiny microbes were able to spread from the carbon-rich birches to fir when they were intimately mixed.

I spent the spring in the trees, living on my own in our Kamloops log house as Don completed his dissertation a thousand kilometers away in Corvallis. If he were here, we could walk through the pine-grass and arnica, sort out where we were going, decide about children. He'd help me remember to turn over the soil in our garden. Clear the papers off the table and clean the kitchen and cook some good food. Instead, I escaped into my experiments, roaming the dry open savannahs bordering the grasslands and the pine forests in the montane. Checking to see who lived, who thrived. Driving the back roads with my hair in tangles and the seat covered in maps and empty coffee cups stuffed with apple cores. Checking in at the switchboard for any phone messages.

Matthew Kelly Charles was born to Tiffany in April. Two weeks later, Kelly Rose Elizabeth arrived to Robyn and Bill, their second child, three years on the heels of Oliver. My new niece and nephew both held my late brother within their names. I sent Matthew a crib and Kelly Rose a lace dress. The days were lengthening, the soil warming, and I was starting to find peace again in being alone.

I returned to my shambolic office one day in June and found a safety citation declaring my piles of journals a fire hazard. Barb appeared, howling with laughter. Under the ticket was a letter from the editor at *Nature*. A critique had been submitted by a lab in England. He wanted me to review it and advise the journal whether it was of sufficient merit to publish.

The first criticism was that the amount of carbon I'd detected traveling through the soil to cedar—one-fifth of that transmitting through the mycorrhizal network between birch and fir—was large enough that it could overshadow the amount traveling through the fungi, negating the mycorrhizal network as a serious transfer pathway. Tapping out the first lines of a response, I explained to Barb that they hadn't noticed the statistical test I'd done showing the amount moving through the soil was not just way smaller than the amount through the fungal network, but significantly so. Besides, I'd made it clear there was more than one pathway for communication.

The second complaint was that the amount of carbon transmitting from fir to birch was so small—one-tenth of that transmitting from birch to fir—that the machines had probably misread the data, and therefore I could not claim a two-way transfer. "We verified two-way transmission in this other case," I said, showing Barb my lab study emulating the field experiment.

The third criticism was that I'd overdosed my seedlings with carbon dioxide when I'd injected the $^{13}C\text{-}CO_2$ into the labeling bags, thereby increasing the photosynthetic rate of the plants and flooding the roots with sugars. If this happened, they argued, more carbon would move to the neighboring plant than would occur naturally. Their complaint arose because I'd used a fair amount of $^{13}C\text{-}CO_2$ so that the mass spectrometer would more easily detect any carbon-13 that might move into the plant tissues. This differed from how I'd plied carbon-14, where a lower pulse of ^{14}C-CO_2 was adequate because the scintillation counter was highly sensitive at detecting the isotope. Barb helped me locate my doctoral lab study showing the CO_2 dose I used in the field had no effect on the allocation of carbon to different parts of the seedlings or amounts transferred.

The final point made me bite my lip so hard it bled. It said I could not claim my seedlings were purely collaborating instead of competing. But I had suggested that the relationships were multifaceted, and that birch was cooperating by sharing carbon even as it competed for light. I hadn't suggested that competition was never involved. They had mischaracterized what I had written, and I was furious that their comments seemed aimed at dismissing my findings. I finished my rebuttal, ending it with my conclusion that the critique was without merit. Barb put it with my other applicable studies into a manila envelope and ran it over to the mailroom. Within a week, *Nature* replied that they had chosen not to publish the critique.

Boy, was that a mistake.

Within a month, I received an email from a colleague who'd heard a keynote talk in Australia by the same lab criticizing my paper. I shrugged this off too, since science is built on peer review. Academics love to pontificate, and I considered myself a scientist much more than an academic. Besides, they were probably confusing the arbuscular mycorrhizal meadows of England, where they'd found no carbon transfer between flowers and grasses, with my rapturous ectomycorrhizal forests, where carbon could move like a luge. No, my colleague insisted; this was a public drubbing. Then another email came from another colleague about a talk in Florida. *Oh dear,* I thought, realizing my naïveté—I should have responded more openly to the critique. Alan had once called publicity a double-edged sword. Don advised me to ignore what I was hearing. Or better yet, publish a reply. He was right, but I couldn't seem to follow either of his suggestions. I convinced myself things would quiet down. I was too tired and guileless to grasp the importance of what was going on, to go public with my responses. The original group soon published an article detailing its complaints.

Soon new journal articles appeared citing my work alongside the rebuttal, giving the complaints equal footing. A shadow was growing over my work. To Don, the solution remained clear. Stop fretting and write. "I know," I'd say, wringing my hands. Dave, seeing I was at a standstill, wrote a rebuttal to the rebuttal and published it in *Trends in Ecology and Evolution.* Others stepped in to help.

It took me a long time to grasp what was happening, but soon I pieced together that I'd waded into a bit of a British scientific debate. There'd been a discourse about whether the carbon Sir David Read had seen move between pines in his lab study amounted to anything in nature, and it had stoked controversies, including ones related to the importance of symbiosis in evolution. The question of whether it was mainly competition that shaped forests was at stake, a long-held assumption based on recognizing that this was central to natural selection. The work with the arbuscular mycorrhizal plants in the English lab suggested carbon transmission through networks was irrelevant. My work, seemingly coming out of nowhere, suggested otherwise. I'd stepped into the middle of a firestorm. In time, I published my rebuttal in two separate papers, but by then my doctoral discovery was already in question.

A few years later, I presented a paper at a conference and, to clear the air, I approached the professor who'd written the original critique. He was engrossed in conversation as I lingered, waiting for an opening. I don't know if he saw me, and I don't know how he could not have, but he did not turn to me. After what seemed like light-years of waiting, I walked away, having come to terms with the fact that this war had less to do with me and more to do with the scientists who'd been battling long before me. I was just a young woman from Canada who had fanned a fire already burning. I knew nothing of their flower-splashed English meadows, and they knew little of my cathedral forests.

But not publishing my own rebuttal within a year of the critiques was a mistake. Among academics, it was like an admission of fault. This hit home each time I read a new article citing my doctoral work with the rebuttals in the next sentences, undermining my contribution. I'd have to recover somehow, stand up. But I was working for the Forest Service, and the importance of my work wasn't clear to their mission, and there was no apparent need or funding for continuing. I gave no talks about the findings to my government colleagues, nor did I discuss the academic debate. Instead, I backed away, ducked under, hid beneath. I wanted children, and I needed time with Don, to be at peace, to learn to like myself again. I needed to grieve. I needed to work on something less fraught, so I turned my attention to other worries of the forest: the increasing insect and disease damage to the trees as the summers and winters grew unusually warmer.

Dr. Melanie Jones, a member of my doctoral committee and also a professor at Okanagan University College, didn't let the matter drop, however. She cared, very much. As coauthor of my doctoral papers, she wanted to address the criticisms, put the debate to rest. She applied for a grant, and with her student Leanne, we repeated my *Nature* experiment, this time including not just one shot of isotopes in the summer but additional doses in the spring and fall, to see whether the direction of net transfer shifted over the seasons. Whether fir gave more to birch in the spring and fall, when fir was growing and birch was leafless, the opposite of what I'd observed in summer.

The first labeling took place in early spring, when the buds on the fir had broken and started to sprout needles but the birch leaves had not yet flushed. At this time, fir was a source of sugar, and birch was the sink. The second labeling took place in midsummer, as with my *Nature* experiment, when the birch leaves were fully expanded and sweet with sugar and the firs grew more slowly in the shade. In this case, we expected to find the same result: carbon moving down the source-sink gradient from birch to fir. The third labeling was in the fall, when fir was still putting on girth and roots while the birch leaves had yellowed and stopped photosynthesizing. Fir was again the source and birch the sink.

Our hunch was right. The way carbon flowed between the trees changed over the growing season. Unlike what happened in the summer, when birch sent more carbon to fir, Douglas fir in spring and fall sent more carbon to birch. This trading system between the two species, shifting with the seasons, suggested that the trees were in a sophisticated exchange pattern, possibly reaching a balance over the course of a year.

Birch was benefiting from fir, just as fir was benefiting from birch. Quid pro quo.

Fir was not draining birch of carbon, but instead was giving it back in the shoulder seasons. The two species were in an alternating feedback system that depended on their size differentials and changes in source-sink status. In this way, they coexisted in harmony. The dynamics of the mycorrhizal network were starting to make sense. By being in place together in a network of fungi and bacteria, birch and fir shared resources, even as they outgrew each other and cast shade. Through this reciprocal alchemy, they remained healthy and productive.

But I still had to test these ideas in actual plantations over the long term. It was important to apply the basic science in realistic settings to help foresters understand how to change their practices. How to combine different species; what distance to space the trees; when to plant, brush, space, and thin. I designed dozens of experiments in mixed forests to illustrate aspects of the dance, how the functioning of the community depends on the terrain, climate, and density of one species versus another, and how much is related to the age and conditions of the trees.

In my experiments, I quantified the different strengths of competitive and cooperative interactions between birch and fir, depending upon whether they were short or tall, young or old. On different types of land—poor or rich, dry or wet. And how they worked together—or against each other—over the long run. This research told me which sizes of trees were most competitive or cooperative or both, and what types of land were most problematic, so that weeding practices could focus on just those elements. In another study, I tested the distance over which birch and fir competed and cooperated, and how this varied with site type, so I could help foresters prescribe treatments that removed only small numbers of birch in local areas around the conifers. In yet another study, I evenly thinned taller birch to different densities and watched how the shorter conifers in the understory reacted.

I tested different ways to treat birch selectively to free up individual conifer trees that were struggling. I compared cutting out individual birches with clippers to poisoning them with herbicides, to girdling them with chains that cut into their bark.

I investigated whether these relationships with birch varied with the identity of the conifer species—whether it was Douglas fir or western larch or western red cedar or spruce—and discovered that they did. Each species cooperated and competed to different degrees and in different ways on various kinds of sites. Knowing the land really mattered.

These experiments are now twenty to thirty years old, but the trees are still youthful, their futures mysterious. In forests, experiments are slow, and the lifespan of a scientist is far shorter. One way to see the future is to use computer models to project how the forest will grow over hundreds of years. To let us glimpse the future, imagine what it might be long after we're gone.

Don had finished his PhD and was back home with me now in the woods of Kamloops. He had rented an office for running his forestry-consulting business, analyzing and forecasting the impacts of different practices on growth. I asked if he'd calculate a model of how productive fir would be after a century of growing alone compared to mixed in with birch. I brought him piles of papers I'd collected over the years, and he scoured them for the data he needed: information on how fast and tall the trees grew, how much biomass they allocated to leaves and branches and boles, how dense the stands became, how much nitrogen the trees stored in their tissues. How fast their leaves photosynthesized, then decayed. He used this information to calibrate his model carefully, slowly tweaking it so it was as realistic a representation of the forest as he could make.

I ran over from my office at the Forest Service when it was time to run the model. He moved a pile of papers off a chair for me as he tapped the keyboard, and green lines of computer code scrolled by and graphs etched out on his screen. "It's as you thought," he said, pointing at histograms proving that clear-cutting and removal of birch was detrimental to the long-term productivity of the forest. The numbers showed that forest growth declined with each successive one-hundred-year cutting-and-weeding cycle. Without the companionship of birch, with its microbes transforming nitrogen along mycorrhizal networks and bacteria helping guard against root disease, the growth of pure Douglas-fir stands declined to half of that evidenced in mixtures with birch. Birch, on the other hand, maintained its productivity without fir. According to this model, it didn't seem to depend on fir for anything. "But I bet it does in some other way," I said, leaning in for a kiss.

IN SPITE OF MY BREAKTHROUGHS—that trees really are dependent on their connection to the soil and to one another—the thing I most wanted was to talk to, commune with, heal with Kelly. I remembered huckleberry picking once in our grandparents' yard when we were little, when he was upset that a bug got into the pail with his two huckleberries. "Get ee out, Gumpa," he'd begged, panicky. I daydreamed about him standing in Grannie's garden, holding the biggest tomato. I pictured us fishing for minnows off the dock with rods we'd made from willow shoots. Slip-sliding over rolling logs on searing summer days in the cold waters of the Arrow Lakes. Canoeing across the North Thompson River to ride Mieko through the corn rows and cottonwoods.

The next spring, I made a garden.

Not any old garden, but one based on the discoveries I had been making when he died. One where the plants could share resources and lean on one another. Where they weren't planted in rows, each isolated from the next, but mixed so they could communicate. Care for one another. I followed the "three-sisters" technique developed by the Native Americans, who grow corn, squash, and beans as companions to enhance the growth of them all.

I'd always made a row for each vegetable in my tiny patch of garden soil. But this year, I built mounds of rich earth spaced about a foot apart, molding each into a bowl as though I were a potter, to keep the water from trickling away, as Grannie Winnie had shown me. I planted a seed of each of the three sisters in every mound and watered them daily, and after a week, tiny cotyledons emerged from the black grains.

Garden plants usually associate with arbuscular mycorrhizal fungi, unlike the ectomycorrhizal fungi on most trees. There are only a couple hundred arbuscular mycorrhizal species worldwide, compared to the thousands of ectomycorrhizal species. These arbuscular mycorrhizal fungi are generalists, meaning that even the few species that exist in nature can colonize the roots and should link most of the garden vegetable plants. Like corn, squash, beans, peas, tomatoes, onions, carrots, eggplant, lettuce, garlic, potatoes, yams.

Within weeks of emergence, the roots of my plants were mycorrhizal, tied together. I pulled up a bean and saw tiny white nodules along its length, housing nitrogen-fixing bacteria. The beans were transforming the nitrogen and adding it to the mound of soil shared with the corn and squash. The corn returned the favor by providing a structure for the beans to climb. The squash served as mulch, keeping the soil moist and the weeds and bugs down.

I imagined how the mycorrhizal network played a part in this dance, my garden's network shuttling nitrogen from the nitrogen-fixing beans to the corn and squash. And the tall, sunny corn transmitting carbon to the beans and squash it was shading. And the squash sending the water it had saved to the thirsty corn and beans.

My garden thrived.

I could feel forgiveness.

I started clearing trails through the forest surrounding our house. Tromping along paths created by animals. Getting to know the shady glades soft with moss, the moist hollows thick with water birches, the grassy slopes where rabbits lived in holes left from rotting roots. The oldest trees and their broods growing in clutches nearby. I'd linger at the pup-tent-sized ant nest, thrumming with thousands of creatures, to watch some crawl single file before I struck out along the main route to the old pine, jumping over the streams carrying needles and lichens.

I thought of new research on hydraulic redistribution by Douglas firs, where the deep-rooted trees lifted water to the soil surface at night and replenished shallow-rooted seedlings so they were vibrant during the day. Had anyone examined whether firs spread water through mycorrhizal networks? Perhaps they shared water to keep their community whole, replenishing their companions through times of hardship.

Plants are attuned to one another's strengths and weaknesses, elegantly giving and taking to attain exquisite balance. A balance that can also be achieved in the simple beauty of a garden. In the complex society of ants. There's grace in complexity, in actions cohering, in sum totals. We can find this in ourselves, in what we do alone, but also in what we enact together. Our own roots and systems interlace and tangle, grow into and away from one another and back again in a million subtle moments.

THE PHONE RANG, and I got up from my kitchen table. I adored our log house, nestled among the Douglas firs and ponderosa pines, the dusty pink roses and yellow balsam root blooming in the meadows. Out of the corner of my eye, I caught the red plume of a pileated woodpecker flicking across the window and landing on a fir branch. It looked at me as I picked up the receiver and listened to a reporter from the Canadian Broadcasting Corporation. Would I do a radio interview tomorrow? The bird cocked his head. I thought about the critique. I was sure they'd ask about that. The woodpecker banged his beak with the strength of a jackhammer, bird and tree needing each other to create a carving. Chips flew, hitting my window. Why should I care so much about the criticisms? I did the work for the sake of the forest, not because of academic hubris. My work was out, and it was time for me to speak.

The tree stood unfazed by the woodpecker's assault, its weathered bark and the bird's beak synchronized like intricate clockwork.

"Yes," I said.

PAINTING ROCKS

November. Snow was blanketing the Rocky Mountains.

I was on a solo backcountry ski trip at Mount Assiniboine, pausing in the pristine cordilleran of Healy Pass. Subalpine firs were bent over in casts of snow and ice, and the whitebark pines were spread-eagled like bouquets of bones, dead from mountain pine beetle and rust caused by the stress of climate change. I was three months pregnant. In our year apart while Don wrote his dissertation, through my long nights after Kelly's death, perhaps because of our loneliness, there came a wordless dawning of the truth that I was thirty-six and he was thirty-nine, and the time for children had come. Skiing into Assiniboine was my celebration of this gift.

The beetles were on a rampage in the couloir. The outbreak had started in Spatsizi Plateau Wilderness Provincial Park to the northwest four years earlier, in 1992, when winter temperatures had increased by a few degrees and the coldest months stopped dropping below minus thirty, allowing the beetle larvae to thrive in the thick phloem of the aging pines. Lodgepole pine had coevolved with the beetles in this landscape, naturally succumbing after about a century to create space for the next generation. As the trees declined, fuel accumulated as a matter of course, and wildfires were ignited by lightning or people. Flames released pine seeds from resinous cones and stimulated aspens to sprout from thousand-year-old root systems, their moist leaves reducing the flammability of the young forest. As fire fingered through the landscape, it petered out in these aspen-clad glades, leaving a mosaic of different-aged forests that was itself resistant to future fires. But in the late 1800s, European settlers disrupted this balance by burning the forest montage to the ground in search of gold, creating a vast blanket of new pine stands whose uniformity was later enhanced by fire suppression and herbicide spraying that ensured the aspens weren't interfering with profits. As these pine trees turned one hundred and the climate warmed, the beetle populations exploded and the landscape ran red like blood flowing through water.

The air rushed cleanly into my lungs as I glided among the dead whitebark pines, intoxicated to be following tracks and carving new turns around rockfalls and tree wells. Don was taking the afternoon to build a cradle. Contentment had enveloped us both. But in the heart of the saddle, I stopped to check some tracks in the fresh snow and felt a familiar rush of fear. The paw prints were as large as saucers, claw marks an inch deep.

Wolves. A lone skier was easy pickings.

I skied away, across the pass. Soon, though, I was lost. When I circled back to the center, I shuddered to be back to my original tracks, already frozen in the drifting snow.

Covered with fresh prints.

Maybe three wolves. Hunting me?

I instinctively kept skiing down the pass. Naked alpine larch, clustered in bowls below the peaks, their golden needles already fallen, were behind me. Down here, the subalpine firs were knotted together in small groves, their numbers increasing as I descended. Telemarking with thirty pounds on my back strained my legs. My baby, no bigger than an ounce of gold, didn't throw off my balance. I cinched up the hip buckle to stabilize myself on the icy, broken terrain and turned slowly, one link at a time.

I made a large traverse to the east to evade a ravine, avoiding a steep section before heading back. It was hard to see because the trees were tightly spaced. Younger lodgepole pines. There must have been a fire a few decades back. Soon I was off course again and checked my compass. If I didn't stay oriented and get back to the main route, this could get dire.

My fear evoked some of my continuing frustration. I had growing evidence that forests have intelligence—that they are perceptive and communicative—but I didn't feel ready to take on the establishment. The fellows would ignore me or, worse, laugh at my talk of the sentience of plants. No, I was pregnant and needed to stay quiet to protect my child, the most precious thing in my life. The CBC radio interview had drawn some interest from local naturalists and environmentalists and even a few like-minded foresters, but it was met with silence from the provincial capital. Without even so much as an email from the policy guys, I wondered whether giving interviews was worth it. Or talks at conferences, for that matter. I couldn't go public more than I already had; there was too much at stake now.

I skied back a hundred meters and found old tracks from earlier skiers. The wolf prints crossed them three times. Now there were at least five animals.

Kelly had lots of stories about wolves accompanying him when he herded cattle.

I skied farther. The lodgepole pine trees grew sparser, and their puffy crowns reached closer to the ground. There should be a special word for the type of mourning you know is to come. In a decade, 18 million hectares of mature pine forest would be dead, representing about one-third of the forested area of British Columbia. The beetles would continue to chew their way through whitebark, western white, and ponderosa pines, through the United States from Oregon to Yellowstone, and would start infesting the jack-pine hybrids across the boreal forest of Canada, producing a total epidemic across North America in an area roughly the size of California, surpassing that of any insect outbreak in recorded history while providing fuel for devastating wildfires down the road. The beetles also infested plantations, especially those with fast-growing pines that had been weeded of their birch and aspen neighbors.

I passed a grove of bare aspens. The prints were melting in steaming urine. Dark orange yellow. I kept to the main route out of the narrow valley, adrenaline making my pack lighter. The wolves stayed in front, just out of sight, leaving only their traces.

Their tracks were heading straight for the main northbound trail, and I suddenly calmed. The wolves were not pursuing me; they were leading me out of the valley. As the vista widened, my trail converged with one from the south. I turned onto it, while the wolf tracks veered abruptly north. A gust of wind blew over them as they disappeared into the trees.

It was as though the wolves had said goodbye.

I lit a candle in the snow for my brother, and for his spirit in those wolves. The lodgepole pines were tall and strong, and their lofty crowns shadowed me while they steadfastly watched over some sub-alpine firs. I needed to linger here where canyon rock and crystallized tree crowns and packs of wolves had come together. The sun climbed over the granite peaks, and I tilted my face toward it. I pulled out my sandwich, ready to stay in this place forever. I felt welcomed, whole. Pure and clean and untroubled.

As I ate, I wondered why trees—these aspens and pines—would support a mycorrhizal fungus that provides carbon (or nitrogen) to a neighboring tree? Sharing with individuals of its own species, especially its own genetic family, seemed like an obvious benefit. Trees disperse most of their seeds—by gravity or wind or the odd bird or squirrel—in their small local area, meaning that many individuals in an immediate neighborhood are related. The pines clustered at the edge of this meadow were probably relatives of the same family, their genes diversified by pollen drifting in from distant fathers. These parent trees shared some of the genes of the trees around them, and sharing carbon to increase the survival of their seedlings, their own offspring, would help ensure the genes got passed to future genera-tions. A later study would show that the roots of at least half of the pines in a stand are grafted together, and the larger trees subsidize the smaller ones with carbon. Blood runs thicker than water. This makes perfect sense from an individual-selection perspective. It's Darwinian.

But my work was showing that some carbon also moved to unre-lated individuals, ones of an entirely different species. From birch to fir, and back again. I looked at the white aspen, its bark basking in the sun, and wondered if it shuffled carbon to the subalpine firs under its crown. And the other way around, firs to aspens. The generalist mycorrhizal fungi might invest in many tree species to hedge its bets for survival, and the off chance that some carbon would move to a stranger was simply part of the cost—collateral damage—of moving carbon to relatives. But this was not what my trees were showing. They were offering me evidence that the pattern of carbon movement was not just by chance, an unfortunate consequence of the moveable feast. No, my trees were demonstrating that they had a lot of skin in the game. Over and over, the experiments showed that carbon moved from a source tree to a sink tree—from a rich to a poor one—and that the trees had some control over where and how much carbon moved.

A squirrel chattered in the branch of a knotty Rocky Mountain juniper, waiting for me to toss it scraps of my sandwich. It kept an eye on the Clark's nutcracker at the top of a pine, probably with a white-bark pine seed in its beak. A raven—a species that also coveted those energy-rich seeds—gurgled a song. Whitebark pine depends on all of these species and more, including grizzly bears, to scatter its heavy seeds. Why would the old pines trust their reproductive success to these birds and animals, whose interest in the seeds was only as food? A few seeds needed to be left to germinate and grow into offspring, to ensure successful reproduction of the elders; why trust that enough would remain? If one of these seed dispersers disappeared, perhaps in a fire or during a particularly harsh winter, then others might deliver the goods. In the same vein, why would a tree pass carbon to a generalist networking fungus—a *Suillus* or *Cortinarius*—that could then pass the carbon to an unrelated tree? From the pine to the understory subalpine fir?

I threw my crust toward the squirrel, and the raven and nutcracker swooped to jockey for the prize. Tail twitching, the squirrel launched off its stump. Just as the old whitebark pines were glad to feed their seeds to birds and squirrels, depending on more than one for dispersal, there must be a similar evolutionary advantage to a tree hosting many mycorrhizal fungal species making up the linking network, letting it benefit from a diverse suite as insurance in case one element got lost.

Maybe even more important was the fungi's ability to reproduce rapidly. Their short life cycle would enable them to adapt to the rapidly changing environment—fire and wind and climate—much faster than the steadfast, long-lived trees could manage. The oldest Rocky Mountain juniper is about 1,500 years old and the oldest whitebark pine around 1,300, in Utah and Idaho, respectively. Meanwhile the trees here would take decades to produce their first cones and seeds and then do so only sporadically thereafter, but their fungal network could spawn mushrooms and spores each time it rained, potentially enabling its genes to recombine several times a year. Maybe the fast-cycling fungi could provide a way for the trees to adjust swiftly to cope with change and uncertainty. Instead of waiting for the next generation of trees to reproduce with more adaptive ways of coping with the soils warming and drying as climate changes, the mycorrhizal fungi with which the trees are in symbiosis could evolve much faster to acquire increasingly tightly bound resources. Perhaps the *Suillus* and *Boletus* and *Cortinarius* fungi could respond more immediately to the warming winters that had spawned the mountain-pine-beetle outbreak and help the trees still gather nutrients and water to maintain a level of resistance.

The raven won the battle for my sandwich crust and spiraled past the Clark's nutcracker in a cloud of feathers and squawks. Not only was the squirrel too slow, it had no hope of wresting anything from the beak of a bird. It would have to unearth whitebark seeds instead, after the birds buried them. Or it could feast on a mushroom left drying in the branches of a pine. It wouldn't live long with neighbors like the raven and nutcracker if it had to rely only on whitebark pine seeds they'd overlooked. Likewise, the fungus could hedge its bets, hitching its spores on legs or feathers, or catching an updraft to colonize new hosts.

If the fungus acquires more carbon from one tree than it requires for its own growth and survival, then it could supply the excess to the other networked trees in need, and in so doing diversify its carbon portfolio—insurance in acquiring essential resources. The fungus could shuttle carbon produced by a rich aspen to a poor pine in the middle of the summer to ensure it had two different healthy hosts—sources of photosynthetic carbon—in case there was a calamity and one died. Like investing in stocks along with bonds in case the market crashed. Then if one of the trees in the network died—perhaps the pine succumbing to mountain pine beetle—well, at least the fungus could rely on the aspen for its energy needs. This more secure carbon source from multiple tree species could increase the survival of the fungus during trying times. The fungus might not care what species the hosts were, as long as at least one of its carbon sources remained viable. Investing in diverse plant communities is a lower-risk strategy than investing in just one species. The more stressful the environ-

ment, the more successful are those fungi able to associate with multiple tree species.

I felt strong and nimble balancing my pack on my hip belt, and I
turned to the fork leading south along Bryant Creek.

Though my puzzling elated me, something still didn't quite fit
together. I thought about the greater group of interacting species. The
whole community of plants, animals, fungi, and bacteria. Individual
selection might explain how the fluorescent pseudomonads interacted
with the mycorrhizal fungi of birch to reduce *Armillaria* root disease
in Douglas fir. *Could selection also operate at the group level?* Individual
species organized into complex community structures that promoted
the fitness of the whole group. Do cooperative guilds of species—like
guilds of people in societies—exist? Where multiple tree species are
linked by a network for mutual aid, in the way it takes a village to raise
a child, despite a risk that there might be cheaters in such guilds. But
this sharing would work if our behavior was ruled by steadfast tit for
tat, like the two-way transfer between birch and fir and their principle
of reciprocity, changing the direction of net transfer over the course
of the summer. Quid pro quo. But what about longer-term shifts in
trade? Such as when fir eventually grows taller than birch. Would the
quid pro quo rule of engagement change, and how might this compare to our human lives when they become more complex and our
relationships transform with age? (If Jean helps me with childcare,
how do I repay her if she moves far away?) I wondered why two tree
species would continue to trade carbon over the longer term, given
the uncertainty of the future.

I thought back to the prisoners in my alder experiment. Because
neither the guard nor supervisor had a weapon, any prisoner could
have escaped. The guy eyeing the timberline sure looked ready to bolt.
One man deciding on his own to run would have amounted to a
betrayal of his fellow prisoners, putting them all at risk of more jail
time. From a purely selfish point of view, the jittery prisoner might
have fled to freedom. On the other hand, if he chose to cooperate
and the others did too, there was the possibility that their sentences
would be reduced for good behavior. But there was no way for them
to know the outcomes, creating the classic *prisoner's dilemma*. It seems
to make more sense to escape, but in the end, a prisoner's instinct is to

cooperate. Studies show time and again that cooperation is commonly chosen in groups, even when betrayal of others could lead to a better individual reward.

Maybe birch and fir, and *Armillaria ostoyae* and fluorescent pseudomonads, are in a prisoner's dilemma where, in the long run, the benefits of group cooperation outweigh the costs of individual prerogatives. Fir can't survive without birch due to high risk of infection from *Armillaria,* and birch can't survive in the long run without fir because too much nitrogen would accumulate in the soil, causing the soil to acidify and birch to decline. In this scenario, the little fluorescent pseudomonad bacteria serve two functions: they produce compounds that inhibit the spread of *Armillaria* root disease among the trees, ensuring there is still a source of carbon energy for the community, and they transform nitrogen using the carbon exuded by the mycorrhizal network. Was this still in line with selection at the individual species level, or was it at the level of the group?

Wolves thrived in a relationship with the forest, and the snow, and the mountains. The animals found food, shelter, and protection for their pups among the trees and interacted with the moose, goats, bears, and whitebark pine to create a diverse community where the players coevolved, learned, and were bound as a whole. Distracted, I practically skied straight into a couple of biologists tracking the radio-collared wolves. They knew the pack well; the leader was an old mother wolf.

I asked why they were tracking the wolves. As the shadows of the peaks grew longer, the lead tracker, a lean, windburned woman with a dark ponytail, told me about the pressure to cull wolves in the park to alleviate the decline in caribou population. She pushed her sunglasses back on her head as she spoke, radiating a fierce intelligence. Her assistant, a young guy with a pack that might throw even Jean, was fiddling with his radio.

"It's the clear-cutting," I said, meeting her gaze. The sprouting willows and alders were attractive browse for moose, causing their populations to increase, attracting wolves. The problem was that the wolves hunting moose also killed mountain caribou, which were in precipitous decline due to habitat loss and interactions with people.

She nodded in agreement as she shifted on her tele skis and checked that her avalanche beacon was on.

"Yes, the snow gets so deep in the clear-cuts the caribou can't out-run the wolves," she said, looking toward the trail where the mother wolf had gone. And there were more and more clear-cuts as the beetle-killed pines were being salvaged.

"Gotta go or we'll lose them," the assistant said, squinting at the tracking device and tightening the chest strap on his pack. The re-searcher's eyes narrowed toward the pass ahead.

"See ya," she said, and I said goodbye too, appreciating her unwav-ering pursuit. They evaporated into the pines as seamlessly as they had appeared, reminding me that a person could easily disappear out here without a trace. It was past noon. I had to keep moving or I'd be skiing the last miles in the dark.

The trail along Bryant Creek was fast and gently downhill, and as I raced past swaying pines, the sun at my back and the avalanche tracks falling behind me, I was grateful the wolf biologists had packed down the trail with their skis. I reached my car as the sky's pink and purple stripes faded to black across the tilted sheets of sedimentary crust.

Ecosystems are so similar to human societies—they're built on rela-tionships. The stronger those are, the more resilient the system. And since our world's systems are composed of individual organisms, they have the capacity to change. We creatures adapt, our genes evolve, and we can learn from experience. A system is ever changing because its parts—the trees and fungi and people—are constantly responding to one another and to the environment. Our success in coevolution—our success as a productive society—is only as good as the strength of these bonds with other individuals and species. Out of the resulting adaptation and evolution emerge behaviors that help us survive, grow, and thrive.

We can think of an ecosystem of wolves, caribou, trees, and fungi creating biodiversity just as an orchestra of woodwind, brass, percus-sion, and string musicians assemble into a symphony. Or our brains, composed of neurons, axons, and neurotransmitters, produce thought and compassion. Or the way brothers and sisters join to overcome a trauma like illness or death, the whole greater than the sum of the parts. The cohesion of biodiversity in a forest, the musicians in an orchestra, the members of a family growing through conversation and feedback, through memories and learning from the past, even if chaotic and unpredictable, leveraging scarce resources to thrive. Through this cohesion, our systems develop into something whole and resilient. They are complex. Self-organizing. They have the hallmarks of *intelligence*. Recognizing that forest ecosystems, like societies, have these elements of intelligence helps us leave behind old notions that they are inert, simple, linear, and predictable. Notions that have helped fuel the justification for rapid exploitation that has risked the future existence of creatures in the forest systems.

The wolves had given me a sign, as had my three-sisters garden, that I could tackle wrongheaded forestry practices. Maybe my child would be fine, even flourish, if I grew bolder. Maybe the hope filling my veins would fill her too.

I felt encouraged by the mother wolf and the biologists tracking her.

I could feel the presence of the pack.

I could feel Kelly covering my back.

I felt a little less worried and afraid, more eager to step forward. Keen to help make the changes my science kept pointing to. I was still getting requests from journalists about my *Nature* article. A woman from Ontario wrote me a letter to thank me for doing "real work for humanity," and another mother concerned about water shortages in California talked of my "message of hope." I sat with these letters in my hands, knowing I had to keep going for my child. For all of the children, the generations to come. I had evidence that could challenge ecological theory, and perhaps also forest policy. I held small seeds of change.

A REPORTER CAUGHT ME in my office months later. I mentioned I was pregnant, expecting my baby any moment, and we joked how easy it was to gain fifty pounds. I was still laughing when she asked me what my discoveries said about herbicide practices. I exclaimed, "Don't print this, but between you and me, for all the good the foresters are doing, they might as well paint rocks." She thanked me and remarked that the article would be out in a couple of days.

Nervous, I waddled to Alan's office to tell him my comment about the painted rocks, and his face dropped. "You can be sure she'll print that," he said gravely.

"But I told her not to," I explained, suddenly weak with regret. A little foot thumped my belly, and I gasped as Alan gestured at me to sit down. He spent the next hour dialing the reporter's number and finally got through to her in Toronto. He explained that printing the comment would rankle the government and could cost me my job. She made no promises. I felt foolish that I'd been careless, but I felt betrayed, too, that as we'd talked of motherhood, she'd landed a comment that could overshadow my message about forest complexity. Even worse, it was terrible to put Alan in the awkward position of trying to head off the calamity.

When Don and I walked that evening along a nearby trail, he tried reassuring me. Newly flushed cottonwoods were closing their leaves for the day. I'd wanted my baby to arrive as the buds burst in the spring, but my due date was already two weeks past, and the Saskatoon bushes were covered with white blossoms. "She's a responsible environmental reporter, I've seen her stuff," he chattered as he threw the neighbor's black lab a stick. I wanted to believe him. "You have more important things to think about," he said. Another shift had been my decision to step out farther with what I was finding—that I wouldn't let it hurt my child, but protecting her also meant giving her a mother ready to fight. We turned back for home at the sunny balsam roots, Don chatting about his parents coming from St. Louis to visit.

A bath that night relaxed my restless legs and emptied my mind. Don built a fire and watched a baseball game, and I went to bed telling myself everything would be fine. At midnight, I woke as muscles tightened like an elastic band across my middle, and I ran my hand across my belly to calm the baby and fell back asleep.

The next morning, at the front door outside our galley kitchen, I crouched down to pick up the morning newspapers and glanced across the meadow of pinegrass tillers to the purple and yellow crocuses I'd planted last fall. I flipped over *The Vancouver Sun*. The headline, "Weed Trees are Crucial to Forest, Research Shows," was followed by my painted-rocks comment within the lede of the story.

Our log walls turned wavy like heat off a sidewalk. Don stared at me, and a flicker bird flew straight into the window. Don leapt up, stuffing down the last of his toast. His eyes darted from my stricken face to the headline. He steered me to the deacon's bench and took the newspaper from me. "It'll blow over," he said.

"I'd like to finish my tea," I said. "Do you think I should finish my tea?"

"Good idea," Don said. He filled the space between us with more reassurances.

When the second contraction came, he grabbed my bag and helped me stand.

Hannah was born twelve hours later.

MISS BIRCH

The painted-rocks comment caused a small earthquake in Victoria, the provincial capital. At least that is what I heard, because I was on maternity leave when the policymakers hit the roof. While they discussed my fate, or so I guessed, I was nurturing Hannah, her bushel of dark hair and archiving eyes traced directly to Don, tying us all together.

A fellow researcher, delighted at my nerve, emailed congratulations and a picture of a pile of painted rocks.

Another colleague sent me a rock he had painted himself.

A maverick postdoctoral fellow invited me to give a seminar at the University of British Columbia because apparently I'd become a bit of a local heroine, though it was the last thing I felt like.

The newspaper article put my job at the Forest Service on the line, and it also caused a resurgence in attention to my *Nature* piece. I was interviewed on *Daybreak* and *Quirks and Quarks* by CBC Radio, and articles appeared in the Victoria *Times Colonist* and Toronto's *Globe and Mail*. When Hannah wasn't sleeping, she was glued to my hip, absorbing my every move as I talked on the phone with reporters, such a literal being-at-my-side that not wishing to disturb her forced me to speak deliberately, concisely, and I grew bolder and fiercer as I juggled the interviews.

In the mornings, I'd feel strangely calm and patient even though I was drained after sleepless nights feeding Hannah. She demanded all of me, and I soon thought little of painted rocks. Don would make Scottish oats for breakfast before heading to his consulting office. With Hannah in a sling, sleeping on my breast, I'd walk the trails for hours, through swatches of spring-green pinegrass and patches of the yellow flowers named butter-and-eggs, and nodding purple and brown chocolate lilies, under the huddles of firs and ponderosa pines and aspens. Somehow, I knew how to do this. I simply could. Each day, I'd see how far I could walk before she'd wake. Sometimes I'd make it all the way to the high grassland, where there was a marsh-clad lake and meadowlarks singing sharp melodies, red-winged blackbirds perching on bulrushes and calling *oh-ka-leee,* and bluebirds nesting in woven pine-needle cups. Home in the afternoon, I'd set Hannah in the shade of the old Douglas fir for a nap, her bassinet no taller than the seedlings that had found a perch there. I'd lean against the enfolded bark and doze with her as the mountain chickadees and pine siskins went about their daily chores in the tangle of a water birch. *Hey-sweetie,* whistled the chickadee, the siskins constantly atwitter—*tit-a-tit*—in flight. The media interviews went well, the commotion died down, and I was left in peace.

Except on one occasion, when Hannah was three months old. I was summoned to defend my research budget before a committee, as were my colleagues from around the province. We'd each have five minutes to justify our funding for the coming year. I had an ambitious list of projects. That morning, I felt like a newborn myself, nervous to be back in public, wondering if there'd be backlash about my press. Hannah was feeding every couple of hours, so I coaxed her into nursing at the back of the lecture hall before my presentation to ensure she'd sleep through it. Barb stood with me in the shadows. The men of the committee sat in the front row, pencils sharpened, pads of yellow paper ready. Right before my time slot, Hannah began to cry, and I fed her once more.

My name was called. Hannah was tightly latched, but I pulled her off as if tearing a wolverine off a moose leg, bundled her into Barb's arms, and rushed down the aisle. Onstage, I started flicking through my slides. Soon the men were gaping, some looking at their feet, others shuffling papers. A calculator clattered to the floor. I glanced at my baggy purple top. Two wet patches were spreading as though from twin fountains. "Oops," I mumbled, my face burning, my smile as tense as a barbed-wire fence, wanting to die on the spot. An older reviewer loudly coughed. My father would have been just as confused or shocked—breastfeeding was not in fashion in their generation. Female colleagues opened their mouths in shared embarrassment. I sped through my slides and ran, Barb on my heels as I escaped out the back. We stood horrified in the sunshine, but Barb, an unflappable mother, burst into laughter that wouldn't let up until I joined her. A month later, I received my budget, lower than my request but enough to continue my work.

I returned from maternity leave when Hannah was eight months, after wrestling with the idea of staying at home full-time. But I was eager to get back to my research, and Don and I depended on my income. Debbie, my caregiver, was reassuring, but the first time I handed over my precious girl—the love of my life in her mauve coveralls, her wrists still bracelets of baby fat, her breath synchronized with mine—Hannah looked at me as though I'd betrayed her. She screamed and clung and sobbed as I ripped her from my chest and closed the door behind me. I stood outside, lungs heaving as I listened to her cries, my world unraveling.

What was I doing? Was leaving my baby with someone else worth it, so I could sit in my government office and stare out the window? Within a week, I felt better. Another week and we'd settled into a routine, and I started to remember my work. I needed to get on with it. As many months ticked by, I sensed more keenly that it was still my duty to explain my discoveries to the policymakers and forest practitioners.

Alan and I returned to his idea of convening a two-day conference and field trip to review the state of knowledge in the province about how broadleaf plants compete with conifers. We'd invite three dozen policymakers, foresters, and scientists to encourage a discussion about the free-to-grow policy and whether the weeding was improving the survival and growth of young trees.

On the first day, I reviewed my slides once more and made Hannah, twenty-four pounds and almost a year and a half by then, an extra big lunch for daycare—three bottles of milk, slices of avocado, cubes of chicken, cheese sticks, and strawberry yogurt. I was jittery and crabby, and Hannah knew something was up. Don dropped her at daycare and me at the college before heading to his office.

Alan opened with a welcome and presentation of the agenda, which involved colleagues showcasing their work on clear-cutting and brushing in different forests, from the rich floodplains on the coast to the slow-growing spruce plantations in the sub-boreal, to the subalpine fir at high elevation and the pines of the Rocky Mountain Trench. I was nervous to see the policymakers from the capital filling the two round tables at the front of the open floor. The regional foresters seated themselves in the next tier, and the scientists scattered themselves farther back, as if to maintain their independence. Alan always said that getting researchers to work on a common goal was like herding cats. I was scheduled to go last, homing in on my research in the local montane ecosystems we'd be seeing on the field trip the next day. Some presentations showed dramatic conifer growth in response to spraying unusually dense swards of thimbleberry and fireweed, but most showed minor increases, or none at all.

Teresa, a sharp, careful researcher from the north, noted in her talk that several aspens could be left without spruce growth reductions on her sites, and that they helped the conifers avoid frost damage. She spoke rapidly, glancing at the policymakers. Rick, a tall, fast-talking forest manager, interrupted to point out a handful of exceptionally large trees in the weeded patches on one of her slides, giants among dozens of smaller ones, as evidence that free-to-grow trees did have the potential to grow exceedingly large, at least in the short term. Dave, the friend who'd achieved his master's and doctorate in tandem with me, spoke up from the back, agreeing with Teresa that complete broadleaf tree removal was unnecessary because only a fraction of the conifers benefited, leaving most still small and even more vulnerable to frost damage than when the aspens were overhead. And this automatic weeding also came at the large cost of reduced biodiversity, which meant that free to grow wasn't a good blanket policy. But he acknowledged it had a good-for-conifers outcome on certain sites in the north where bluejoint grasses invaded following logging.

When my turn came, I showed data from several experiments, explaining how many plant species—the kind usually targeted by weeding programs—were not hurting the planted conifers as much as expected, if at all. On most logged sites, the conifers grew as well among the native plants—fireweed, pinegrass, willow—as they did when the plants had been removed. The effect of birch on fir was complex and depended on how dense the stands were, how rich the soil was, how the site had been prepared, how good the planting stock was, how much *Armillaria* root disease was present in the original forest. Responses depended on the conditions and history specific to each site and required an understanding of a local forest. I showed data on how many birches could be left in particular situations to ensure good conifer growth while minimizing root disease and maintaining biodiversity. My research was rigorous, but it was also as young as I was. My colleagues nodded where my findings aligned with theirs. I continued on to my last slides, feeling optimistic.

Shrubs such as alder and soapberry, I explained, were beneficial to their needle-leaved neighbors because of their ability to host symbiotic bacteria that fix nitrogen. Not to mention, I thought to myself, their role in providing food for birds and medicine for people and carbon for the soil. In preventing erosion and fire and disease. Toward making the forest a lovely place to be. The policymakers at the front table watched silently at first, but then I registered frowns, and I was further unnerved when a senior manager in his sixties interrupted me to say, "Your data is too new to prove the plants are not overtaking the conifers."

A young forester at the next table, a green baseball cap shading his eyes, called out that my research didn't jive with what the plants were doing in *his* forests. He glanced sideways to some of the older guys for their approval. The Reverend, silent so far, sat immobile as the others at his table stacked their papers, ready to call it a night. *That's okay for now,* I thought. I finished my talk, Alan thanked everyone, and the scientists looked ready for a beer. The policy guys got up together, talking of regulations before relaxing and following Dave and Teresa to Duffy's Pub. My solace was overhearing a forester who'd been quietly taking notes say to a friend, "Well, that was useful. I don't want to brush where I don't have to."

Don and Hannah in her car seat were waiting. She squealed as I kissed her before I flopped next to Don, threw back my head, and groaned, "Oh, God. The other researchers had good data, but the policy guys were still skeptical of my results," I said.

Don, always more optimistic, assured me things would go better when we were all out together looking at the trees.

The next day, I planned to present three sample Douglas-fir plantations that showed a range of conditions—"the good, the bad, and the ugly," representing the natural variability of birch that seeded into clear-cuts across the landscape. One was a representation of the vast majority of plantations—where a low density of birch had seeded in or sprouted up on its own after clear-cutting. The other two plantations were examples of the rarities, where an abundance of seeds had found a foothold and sprouted up in thickets, or where next to none had established and sprouts were scarce as hens' teeth. The plantations were young, about ten years old, the age at which weeding was normally conducted to meet free to grow. I'd chosen these sites to convey that birch was not usually as competitive as the policy was supposing, and foresters were therefore prescribing interventions that did not agree with local conditions. Overestimating the threat of a few birch neighbors could bring unexpected consequences, potentially setting the forest up for a vulnerable future, where lowered biodiversity might reduce productivity, increase risk of poor health, and augment the spread of fire. What we do in these early years of development, after all, determines future resilience. Just as it does with children.

By presenting my argument directly in the forest, in the midst of trees, I thought we could more easily come to an agreement that the policy needed adjustment to better reflect what was happening in nature. Because love of the forests was the one thing we all had in common. Alan and I had rented a shiny Suburban for the day, and we led the cavalcade of vehicles north along the river from Kamloops, with forest-manager Rick and the Reverend in our backseat. Jean and Barb were bringing up the rear in our field truck. Alan, a gracious host, talked easily about the rate of provincial harvest and the backlog of inadequately reforested clear-cuts, and everyone debated who might head the next research funding initiative, but I stayed quiet. Besides, I was queasy, a few months into my second pregnancy. I pretended to look at the maps and notes. Rick chattered with a rolling laugh, describing his favorite experiment up north where the grasses choked his spruces, a reference point for his policies. As we sped past sandbars of black cottonwoods and scree slopes of Douglas firs, the Reverend talked about thinning forests that exceeded a specific density, one that he and the modelers thought would be deleterious to the trees, and this was how they could create more uniform forests that would grow faster and more predictably. I wasn't capable of penetrating the conversation; I'd let the forest speak for me.

Just before East Barriere Lake, we stopped our line of vehicles at a century-old stand of Douglas fir and paper birch. I hadn't thought of myself as the trickster coyote, as the whistler in the song, but I was already worrying that this field trip might cement me as a rebel.

But the old forest felt peaceful and forgiving as I stood on a hummock in the midst of the firs—about thirty-five meters tall—and the shorter paper birches, their leafy branches reaching into canopy gaps. Several cohorts of the elder firs' offspring nestled in the openings. The men jostled and joked and sipped their coffee. Pointing to a sapsucker, Teresa slipped into a conversation with Rick about cavity-nesting birds. Alan stood bandy-legged next to another prime policy guy, and they talked about how the high-density spruce plantations of Scotland should be converted back to native oak woodlands to improve the habitat for birds. Alan, always seeking common threads, pointed to an owl in a birch cavity and commented that birch here were like the oaks of the British Isles. There was little of the tension of the previous day, though the Reverend grumbled about the cold. Jean and Barb had their clippers at the ready to clear the trails ahead of the group.

"I want to start by pointing out that our data are showing that these mixed forests are producing more total wood volume than pure conifer forests," I said. "Even though there is less fir volume here than in a pure-fir stand, these individual firs are growing faster, and when you add the birch volume on top of the fir volume, the total amount of wood on this site is about a quarter more than in a pure-fir forest. It's partly because birch provides a lot of nitrogen to the coniferous trees, which they're short of. They also protect the firs against *Armillaria* root disease, which slows the trees down if it doesn't outright kill them."

Rick said, "Well, that may be true, but let's face it, the birch here

has no value in the market." A nerve in my neck twitched. Not recalling the pleasant chat about owls and their housing needs, the Reverend added that most of the old birch was rotten anyway. Teresa
and Dave stood silently, knowing the current market value for birch
two-by-fours was low, and these birches did have a lot of rot in them.

"You mean the old market," Alan said, jumping in as if he'd been
poised on the end of a diving board. "Markets are changing, and
birch will eventually become more valuable." My arms hung loose as
I breathed in his confidence. "It grows here so easily—it doesn't make
sense to stop what wants to grow naturally and spend a lot of money
doing it. It would be better to cultivate markets for birch products
instead. Then we can build cottage industries to make birch flooring and furniture instead of importing them from Sweden. Look at
lodgepole pine—twenty years ago we called it a weed, and now it's one
of our most lucrative commercial species." The wind rustled through
the pathfinder plants, their leaves gently bent forward in pale green
arrows.

"No one will buy our birch, though," Rick said. "It's too old and
rotten and crooked to run through our mills, and we can't compete
with the Swedish birch dominating the market."

"That's true," I responded, knowing he was right. "But I've been
doing experiments where we've thinned birch saplings to different
densities. We look at each stem individually and select the straightest ones to keep. We remove the decaying and crooked ones instead
of letting them thin out themselves. If we take care of the stands like
that, we can grow a straight, solid birch in a quarter the time of a
conifer."

"But it costs too much to haul the old birch out of the bush," said
the young forester with the green ball cap. That was why it got left on
the landings to rot after the conifers were cut. Teresa nodded, and I
knew it was true, but I wanted us to work through this, to talk about
how we could use some of the old stems and at the same time cultivate the naturally regenerating birch while keeping the stands healthy.
Why was the Reverend so quiet?

"Maybe the government can provide incentives," suggested Alan.
"The companies could have the old birches for free, with no payment
to the Crown, and we could allow young birches as crop trees in new

plantations, managing them using the selection techniques Suzanne has been working on." Alan picked up a chunk of birch a firewood cutter had left and handed it to the Reverend to show the value even now of the wood, and Dave toed a chanterelle, saying that people living out here depended on birch in ways invisible to the government.

"We already have markets for conifers," argued the Reverend, his first comment of the afternoon, as he looked at the piece of firewood, then dropped it.

A studious, sensitive expert in pathogens turned over a birch log with a honey-colored mushroom growing from it and peeled the papery bark to reveal the wood inside, soft, crumbly, and moist. He plucked the mushroom and pointed to the luminescent mycelium infecting the pulpy wood. The men crowded around. As birch reaches about fifty, near the end of its lifespan, it becomes more susceptible to *Armillaria sinapina,* with many risking infection of stem and roots. *Armillaria sinapina* was like *Armillaria ostoyae,* but it infected mainly broadleaf trees like birch rather than conifers. Both fungal species occurred naturally in these forests, and both facilitated natural succession and increased the heterogeneity of the woods by killing trees, opening space for other species to augment the diversity. But *Armillaria ostoyae* was viewed as a bad fungus by foresters because it particularly killed the fast-growing conifer coveted in the marketplace. Weeding the clear-cuts of birch and aspen made the situation even worse because the new stumps provided a rich food base for this fungus to grow, increasing its potential to infect the planted conifer seedlings. Killing birch also reduced the ability of the conifers to resist infection because of the loss of beneficial microbes. *Armillaria sinapina,* on the other hand, was of less concern because it didn't usually infect the coniferous crop trees. It eventually killed the birch trees, though. As the aging birches filled with decay, their leaves yellowed, branches dropped, and insects and other fungi moved in to feast on the sloughing sugars. Sapsuckers and woodpeckers fed on the insects, and when finding a perfect spot, made cavities in the wood for laying their eggs. The long-lived conifers reached into the new space and commandeered the rays and raindrops and sopped up the released nutrients. "The fungus kills the birch, and the gaps become homes for other species, adding to the diversity. It's the natural succession of these forests," the pathologist said, the men murmuring in appreciation.

"When it's young, though, birch photosynthesizes at higher rates than the conifers, and sends more sugars to its roots, and eventually large amounts are stored in the soil. If we start managing forests for increased carbon storage—to slow climate change—birch might be a good choice," I continued. A goldfinch clung to a speckled birch twig and pecked at the gyros of seeds, some trickling to the forest floor.

"Climate change? We don't have to worry about that too," said someone else. It was true that there were so many unknowns about global shifts that we'd been slow to link the beetle outbreak with warming winter temperatures. With so much uncertainty, the government did not have a mandate to take this new drumbeat about the climate seriously.

"Well, the EPA thinks we should," I said, surprised at how assured I sounded. "I've seen the projections, and climate change is going to be our biggest threat before long. We're going to need birch and aspen to grow quickly and put a lot more carbon in the soil, where it will be safe from fire." I went on to explain that, in most years in Canada, more carbon is lost to wildfire than from the burning of fossil fuels, and we should be trying to reduce the risk by planning for landscapes of mixed forests instead of coniferous forests, and for corridors of birch and aspen to serve as firebreaks, because their leaves were moister and less resinous than those of conifers.

"Climate change just isn't happening here," the forester with the ball cap argued. "Look, this has been the coldest, wettest summer ever."

"I know, it's hard to believe when we can't feel it. But the climate models will surprise you," I said, sweeping my hand in the shape of a hockey stick to trace a graph of how atmospheric carbon-dioxide concentrations have skyrocketed since the 1950s.

"You're a birch lover," the green-capped fellow piped up.

"Yeah, I guess I am." I laughed awkwardly.

"We should move on," the Reverend suggested. He whispered something at Rick. When they turned to leave, the others followed like flocking birds, and I zipped my sweater in the chilly wind.

The ball-capped guy asked if he could take my seat in the Suburban carrying the policy guys. I could have kicked myself for saying yes too eagerly so I could jump in with Jean and Barb, hoping Alan wouldn't mind my abandonment. "You're doing great," Jean said, patting me on the arm but looking uncertain.

"This is going to be tough," Barb said as she led the pack down the road.

"They're going to go crazy over the birch at the first plantation," I agreed, feeling heat spread through my nerves like a ground fire through grass.

But these guys knew these sites existed, so we needed to talk about them.

We rolled up to the dense stand of paper birch with some scraggly firs underneath, what I called "the ugly." The site had been mismanaged from the get-go. The loggers clearing out the birch had torn up the ground so badly that it ironically had turned into a perfect seedbed for the tiny winged seeds flying around in the late fall. Then the forester in charge of the replanting had prescribed fir seedlings better adapted to a more southern climate—a perfect storm of planted firs being doomed and the "weed" birch reasserting itself. Now the birches were three meters tall and the planted firs, unequipped to handle the frost, were almost dead. It was by all means an extreme case of birch winning the race. But there were two parts to this stop, and the second portion would bring home the point I wanted to make. Across the road, they'd cut down all of the birches so the firs were free to grow, but the firs were still small and yellow, showing that killing birch to meet the policy goals did not solve the problem.

As we walked into the dense part, I realized my idea was misguided— the whole field trip was heading toward fiasco.

"You see? This obviously shows birch kills conifers," muttered Rick as he found a suffering Douglas-fir seedling. The forester with the green cap looked almost giddy.

"My models plotting growth against light would predict that this Douglas fir will be dead in a couple of years," said Dave, whom I had come to love over the years, and who was just speaking honestly about his data. But he was bringing it up before we'd had a chance to traipse to the other side of the road to see the fir that would soon be dead too, even with the birch gone. I could have throttled him.

"Yes, but my point is that these kinds of stands are rare," I countered, leading them across the road to where all the birch had been cut down. Removing the birch had made not one bit of difference to the health of the firs—they were sick from being planted in the wrong place. "We can easily avoid creating stands like this. We can do it by planting better trees and timing site preparation so it doesn't coincide with seed dispersal of the birches. We'll look at sites where we got completely different results with better site prep and choice of planting stock." I was on edge but had designed the field trip so that the solution would be clear at the end.

We pushed on to "the bad," a plantation where the birches had been cut to the quick and the stumps soaked with herbicide to attain free-to-grow status. The monoculture of firs stood out against the mountainside of birches and cedar like a lawn in a prairie. Jean ran to where she'd painted the dead birch stumps blue, as if she'd sprinkled confetti, and pointed to some planted firs tinged yellow from root disease. Some firs were in better shape, but a tenth were completely dead, skeletons of scratchy, gray twigs. When the birch had been clipped, the *Armillaria ostoyae* fungus had infected the stressed roots and spread to the roots of the intermingled firs. Douglas fir, lodgepole pine, and western larch were heavily favored for planting but paradoxically were the most vulnerable to this type of infection. Rick and the Reverend walked past the ailing firs and pointed to the foot-long leaders of some healthier ones, saying that disease wouldn't strike most plantations. The pathologist said, "There's no *Armillaria* past fifty-two," waving in the direction that lichens were crusting the birch bark, meaning the disease was not a problem for the northern half of the province, where Rick had set his compass.

I was on a raft that had sprung a leak.

Alan handed out colored graphs to show that height growth of firs in one of his species trials was double the ones here despite leaving the birch uncut. As they examined the colorful lines, Alan looked to me to take the mantle. I talked about the *Bacillus* bacteria on birch roots that fix nitrogen, and the fluorescent ones that produce antibiotics and reduce the pathogenic infections on nearby firs. Leaving a healthy mix of birches with its helpful bacteria, I argued, could enhance the health of the fir, like a public immunization program. "The bacteria are fueled by carbon leaked by the mycorrhizal networks as carbon passes back and forth between birch and fir," I managed to say, distracted by the green-capped forester's snickering, but I kept on, adding, "We can surgically remove a few birches to release the firs but retain most of the birch to keep the infections down."

Rick, veering into the center of the group, interjected that the best way to reduce *Armillaria* root disease, according to a study started in 1968, was to pull the infected tree stumps out of the ground after clear-cutting, then plant the firs. I'd been in the field with him before, just us looking at plantations, and he'd been eager to discuss brushing, making a point of citing literature, which I'd found odd since he seemed more focused on that than on looking at the actual trees. I fought irritation. He was right that destumping was standard practice, and there was ample proof it worked at reducing disease. But we needed to find alternatives, I explained, because destumping compacted the soil and destroyed native plants and microbes. "And it's expensive," I said.

"Yes, but it's the most reliable treatment," the pathologist said, pounding in a final nail.

Sounds of agreement arose like croaking, and I felt the stress hormones bathing Hannah's unborn sister.

By the time we got to "the good," where a lovely mix of fir and birch grew in perfect balance, Rick had run out of patience. I didn't have a chance to explain that this parcel showed how birch and fir helped each other, that they were in intricate balance, and we only needed to be patient to let them play out their two-step through the seasons and years. He was angry, and the mood of the policy group had soured.

Maybe he thought my science was bad, or he was starting to see a crack in his policies. To be sure, selective weeding was needed in some cases, but in most plantations, wholesale broadleaf tree removal was simply not justified. But he wasn't about to let me upset his applecart. He moved to within inches of me. I instinctively put an arm over my waist while noticing how terrifically tall he was. I scanned the bushes for the others, but they were scattered. Alan was out of earshot, talking with Dave. Foresters are always looking at some tree or another, or buds and bark and needles. Barb and Jean were by a graceful birch, frozen.

"Well, Miss Birch," he said, "you think you're an expert?"

I'd heard this name whispered behind my back. Birch was the clever substitute in public for what some of them called me in private.

Then he became furious. "You have no idea how these forests work!"

My baby stirred for the first time, and I felt faint.

"You're naïve to think we're going to leave these weeds out here to kill the trees!" he roared.

I opened my mouth, but no words came out. A black-capped chickadee fluffed her wings inside a crown of birches. Three tiny yellow beaks opened like clamshells around her, but their songs for food were muted. The awful things I'd heard about women speaking their minds—comments made even in my own family—echoed in me. The criticisms doused on women behind their backs, even if said in jest, always burned my ears. My Grannie Winnie was quiet, but in large part her resorting to silences to avoid barbs was likely because it was— easier. I'd vowed not to provoke the criticisms of the men, and yet here I was. Barb's eyes were as wide as full moons, and Jean seemed ready to scream.

The men encircled me, nearer than the wolves when I was lost, and I stepped back.

Alan appeared by my side. "Time to go, fellas," he said. Barb hurried to me and said under her breath, "Sheesh." I wanted to crawl away like a beaten dog.

Chickadee-dee-dee sang the chickadee—all clear. The field trip was over.

That evening, I drove Dave to the airport, and we managed to talk about our kids, his cabin at Hudson Bay Mountain, and the upcoming salmon run in the Skeena River. It took an hour to curve down from the dense mixed cedar-birch-fir forests in the mountains and gather speed through the dry, open Douglas-fir forest along the river. I wondered what the mycorrhizal network looked like under these two different canopies. In the thick, moist forest where the trees were of different species but the same age—having regenerated after an intense blaze that killed all of the old trees—I imagined a brilliantly complex network, with hundreds of host-specific and host-generalist fungi, some that linked trees of different species and others that linked trees of the same species. As the forest opened up in the dry valley, where there was only Douglas fir, where frequent understory fires created openings for seed that was spread by old, thick-barked survivors, leading to periodic surges of fir regeneration, I wondered how its underground map would compare. The old-growth trees in this arid landscape seemed to be aiding the establishment of new seedlings, but maybe the mycorrhizal network played a role too in this facilitation. With the fungus serving as a pipeline for carbon, and maybe water, from old to young in the dry soil, as it did from birch to fir in the wet forest of my doctoral research.

The dry forest seemed like the perfect place to make a map of the belowground network, because of the likelihood that connections between trees of the same species would be much higher than between the diverse suite of tree species in the wetter mixed forest. Here in mostly pure Douglas-fir forest, the mycorrhizal fungal community should be dominated by Douglas-fir-specific fungi like *Rhizopogon,* providing it with an exclusive, highly coevolved partnership, where the Douglas-fir seedlings should be connected to the old firs by that single fungal species—like moons in the orbit of planets. A network comprised of a single host-specific fungal species connecting a single tree species, after all, should be more straightforward to map than a network made of multiple host-generalist fungi connecting multiple tree species. Maybe one day I could make a map of the dry fir forest— simple, sharp, and clear—as an easier place to start than in the mixed forests where I'd traced carbon transferring between birch and fir.

Dave offered to help me edit one of my manuscripts that had been rejected by a journal. One reviewer had written, "We can't publish articles by people who think they can just dance through the forest looking at trees." The comment hurt, but I was getting better at taking such dismissive criticisms with a grain of salt. Eventually we reached the landing strip among the bunchgrasses and buttercups at the east end of Kamloops Lake. Dave's eyes darted from the check-in counter in the airport to the orange vinyl seats in the waiting room to the baggage area, and he laughed that the building was even smaller than the airport in Smithers where he lived.

We were sharing muffins near a window reflecting our dusty images when he blurted, "I talked with Rick about what happened today. I told him you were one of the best researchers the Forest Service has."

I tried to hide that I was on the verge of tears. "What did he say?" I asked, not really wanting to know.

"He didn't agree." Dave looked straight at me, but I watched a cowboy ordering coffee.

"At least he's honest," I said, laughing.

"I don't get why these guys are so worked up about you," he said.

I didn't know either. Maybe they didn't like criticism. Or they couldn't listen to women. Undoubtedly they were still angry over the painted-rocks comment. When Dave's flight was called, he wrapped me in a bear hug and disappeared.

To make matters worse, a letter had been placed in my personnel file at the Forest Service in reprimand for the painted-rocks interview. One manager said I could be disbarred by the professional regulators, the Association of British Columbia Forest Professionals, for speaking against government policy, which to him was an example of ethical misconduct. The government foresters increased their scrutiny of my research, and those in charge had my peers review one of my articles even after it was published. I started to feel excluded from new initiatives. My research seemed to be grinding to a halt. On one occasion, when they were threatening to pull funding earmarked for publishing one of my reports, Alan convened a conference call with the policy-makers, and I joined him on a speakerphone to explain that I was requesting only enough to publish my results on the effectiveness of brushing in this region.

"It's not the cost that's the problem. It's the results you're reporting," was said into the air.

"But my results have been through thorough peer review, not just in government but from outside scientists," I said, my voice straining. Alan presented the case that spending ten thousand dollars to get the results out there was well worth it and small in comparison to the hundreds of thousands already invested in the decade of fieldwork. He was steadfast and persistent, and in the end, they reluctantly funded the publication of my report.

Every night as I fought these battles, my growing belly pressed against Hannah's crib as I watched her sleep, I wondered how everything had come to this, the frustration and humiliation at being hauled out in front of my colleagues. I deeply loved the forest, was proud of my work, and yet I had been labeled a troublemaker.

The scientific community was also suspicious. The belief that competition was the only plant-to-plant interaction that mattered was so strong that when I submitted manuscripts for publication, it felt as if my experiments were being picked apart, bit by bit, for mistakes that weren't there. Maybe this was how things were done, and I was inexperienced. But I couldn't help but think they resented that I had published my findings in *Nature,* whistling past famous scientists who had been trying to unravel the mystery of how networks influenced plant-to-plant interactions for some time.

Five months after the field trip, my daughter Nava was born, instantly looking around in amazement at all the commotion. I packed Nava in the baby carrier and put Hannah in a backpack, and we would pick up Jean and hike through the savannah forest, looking for blue jays and cactus flowers. My 398-page book-length report that

Nava (left, one year old) and Hannah (three) in front of our log house, 2001

had been treated like a dog's breakfast was published, and the thousand copies flew off the shelf. A forester later showed me his copy, the cover worn ragged and his favorite pages marked with colorful tags, and he told me it was his bible.

When Nava turned eight months, I returned to work, but Alan could see the writing on the wall. He encouraged me to look for a new position. The conservative political party newly in charge was also downsizing the civil service along with scientific innovation, so scientists were being told to leave if they could.

My maverick postdoc friend at the University of British Columbia, now a professor, soon contacted me about a new professorship opening up. I'd never imagined being in a tenure-track position at a university, but a member of the search committee traveled to Kamloops to talk specifics, and I was encouraged to publish a few more articles to better position myself for the competition. By then I was already dead tired, with Hannah now three and Nava one. Nava was weaned but still at my hip, and Hannah was as rambunctious as a puppy. And I loved our house in the forest and our evening walks along the woodland trails and my hundreds of experiments I'd nurtured like my own children. Besides, I was forty-one—wasn't I too old to start a professorship?

I applied anyway. Don agreed I should try but said he didn't want to move to Vancouver, even though he didn't like living in Kamloops, where I had my government job. Ever since he'd laid eyes on the little town of Nelson, nestled in the Columbia basin not far from Nakusp where my mum grew up, he'd wanted to move there. Its forest was lush, the town small, the pace slow, the people educated, liberal, and artistic. I understood—the draw was strong. After all, most of my immediate family now lived in Nelson, and our girls could be close to Mum—their Grannie Junebug—and Auntie Robyn and Uncle Bill and cousins Kelly Rose and Oliver and Matthew Kelly. But it was so small and remote that there was no work for us there. And—unfathomable—I would no longer be able to pursue my research. Short-listed from a hundred applicants by the search committee, I flew to Vancouver in the dead of winter for an interview, telling myself I could take it or leave it.

A number of months later, Don, the girls, and I were visiting Mum in Nelson, the snow barely off the passes, the ice on the lake only recently thawed. The first sailboats were jibing by on Kootenay Lake, and the leaves of the snowberry bushes along the shoulders of the tree-lined streets were unfolding, and Don breathed out a wistful sigh. As we drove up Kokanee Avenue to Grannie Junebug's house, Hannah shrieked with delight about Easter-egg hunting with her cousins, and Nava laughed alongside her, even though she'd just turned two and didn't know what all the excitement was about. Grannie was at the door, crayons and coloring books in hand. A kitty with fluffy gray fur and six toes on each paw, named Fiddlepuff, pounced at butter-flies skimming the lawn. Hannah ran up the stairs, Nava in tow, Fid-dlepuff behind, and I opened my laptop to find an email from the university offering me the job.

Mum instantly said I should take it. Suddenly it was real, and I was enticed and flattered and rejuvenated. But Don reminded me about what he'd been saying. He'd escaped his native St. Louis and wasn't keen on living again next to factories and bakeries, freeways and sub-ways, jammed-together homes and skyscrapers, where the closest trees were in city parks. But I pointed out that since I was going to lose my job, and since he wasn't wild about living in Kamloops, maybe the big city was the adventure we needed for a while. It would solve our looming financial uncertainty.

We stood under Mum's apple tree, the girls inside with Grannie while Don and I argued to the tune of his refrain about "no intention of living in Vancouver." He waved a hand toward Kokanee Glacier, where we could hike and ski, and said this was why he had wanted to come to Canada. "Just be confident in who you are, and you won't need this job," he explained. "Between the two of us, we can make a go of it here."

I looked toward the mountains, where the cedars cast shade on the devil's club and skunk cabbage, where the sweet organic smell of the forest floor rushed up your nose, where the fresh, tumbling water made your hair soft, where the huckleberries grew on stumps and the wild ginger bloomed in trickles. Where the ancient forests were gradually being clear-cut and planted with rows of firs and pines and spruces.

"But I'll never get an opportunity like this again," I said, a vision of the offer swirling round and round before disappearing down a drain. Don wanted a laid-back life, away from expectations of being a doctor or a lawyer or an accountant, close to a ski hill. "Meet my son the doctor," his mother and aunts would say about his brother and cousins, while Don and his dad talked of fishing and baseball. Even when he was twenty-nine, when I'd met him, he'd talked about retiring to the mountains, but I was so engrossed in my quest to understand the forest that I hadn't taken it seriously, not a clue that it was more than just talk.

I peeled back one of the three-pronged bracts of an open fir cone and ran my finger over the red heart-shaped indentation where the winged seed once lay. Mum's garden bed had a new fir seedling, the seed coat having fallen from the cotyledons. The bark of this little tree wouldn't be thick with wrinkles for another hundred years.

"I love Nelson too," I said. But I wanted the professorship, because I wasn't going to have a job much longer. No matter what we decided, one of us would be unhappy. And what if I couldn't handle the work? The city might be as awful as Don feared. And I worried about putting too much strain on our daughters, our marriage.

"We don't need much money. We can just live in the woods," Don said. I looked past the steep roofline of Mum's yellow two-story Victorian, designed to shed slabs of snow, and across the alley to the neighbor's yard, and I worried that he could hear. Don's voice seemed loud.

"But what about my work? I still have lots of questions," I said, throwing the cone into the flower patch as if pitching a ball.

"Suze, Nelson is a better place to bring up kids," he said, his lip twitching, something I'd seen only once, when we'd argued over whether to go back to graduate school.

We went out for dinner at the fancy All Seasons Café. I ordered sockeye salmon. Don ordered something vegetarian, and we avoided eye contact until I said, "Just think of the fun we could have with the girls."

He pushed his plate to the side and stared squarely at me. "I know exactly how it will be. It'll take two hours to drive through the city to get to the forest, and when we get to the peaceful hiking spot we'd been dreaming of, a million other people will already be there." I didn't know what he meant. When I'd lived in Vancouver as an under-

Douglas-fir Mother Tree, a century old, in the coastal rain forest of southern British Columbia. The neighboring trees are Douglas fir, western hemlock, and western red cedar, and the understory is rich with sword fern and red huckleberry. Sword fern leaves are used by the Aboriginal people of the Pacific Northwest as a protective layer in pit ovens, for wrapping stored food, and as flooring and bedding. The rhizomes belowground are dug in the spring and roasted, peeled, and eaten. The red berries of the huckleberry are used as fish bait in streams, dried and mashed into cakes, and juiced for an appetite stimulant or a mouthwash.

Sitka spruce Mother Tree among western hemlock and spruce saplings along the Yakoun River on Haida Gwaii. Some of the saplings are regenerating on decomposing nurse logs, which protect them from predators, pathogens, and drought. The Haida, Tlingit, Tsimshian, and other West Coast peoples harvest the spruce roots to make water-tight hats and baskets, and they eat the inner bark or dry it into cakes to eat with berries. The raw young shoots are an excellent source of vitamin C.

Suillus lakei mushroom with white fungal mycelia emanating from the base of the stipe (stem). The mushroom fruits from the fungal threads that spread through the forest floor and connect with the nearby trees. The trees provide the fungus with sugars from photosynthesis in return for nutrients the fungus collects from the soil.

Ectomycorrhizal root tips with abundant emanating fungal hyphae. This photo was taken in a minirhizotron at the Oak Ridge National Laboratory.

Mother black bear with her two cubs

Bald eagle

Western red cedar

Fungal rhizomorphs reaching from a mat of ectomycorrhizal fungus

Ectomycorrhizal fungal network in the upper horizons of a soil profile

Thousand-year-old western red cedar Mother Tree in Stanley Park, Vancouver, British Columbia. The vertical scar is from traditional bark stripping, and hence this is known as a culturally modified tree, or CMT.

Sitting against a western red cedar Mother Tree

graduate, I'd never experienced crowds like that when I went hiking or skiing.

"It isn't that bad."

"There was no wilderness like this in St. Louis."

"We can visit Nelson in the summers."

"I won't be Mr. Mom," Don said, and the fellow at the next table looked over.

"I'll be right there, you won't have to do it all," I said, my voice straining to stay low.

"No, I know what these academic positions are like. I watched the professors at Oregon State working their lives away. I know you, and you'll be at it nonstop, and I'll be left looking after the kids because I'm not sure I can find enough work there," he said. Don's niche in data modeling and analysis was small, the client pool highly special-ized, and he knew almost no one in Vancouver. His other choice was to work for a bigger consulting firm, but he didn't like the thought of reporting to someone else after so many years of working indepen-dently. His interest in bush work had always been less intense than mine, maybe precisely because he was from the city. Or maybe he was more interested in building things on his computer or in his work-shop at home. Whatever it was, at this moment we seemed from dif-ferent planets.

The next day, we looked at land for sale above the Kootenay River outside Nelson, where a couple had made a clearing in the woods, with a rolling aspect looking over the river, the needles of the skyward larches brilliant green, the forty-meter-tall crowns of the firs dark and lusty. A stroller sat on the landing cleared for a future house, and a young straw-haired woman emerged from the tent with a baby on her hip and a toddler in hand. They'd tried to make a home, but the woman had given up because the tent had no heat or running water. Her husband invited us to walk the property. I pulled Hannah and Nava over logs and through bushes, and we sat under the larches. Don talked dollars with the fellow, and I thought about how beautiful but impossible this was. We would spend all our time chopping wood and growing a garden, and *neither of us had jobs*. We continued to argue about ways of life, money, what each scenario meant as we took the kids to Lakeside Park and walked down Baker Street, looking at art and books, and we bought ice cream at the counter at Wait's News, where Grannie Winnie had bought us ice cream when I was a kid decades earlier.

A few days later, sitting with the girls under the apple tree, Don said, "Okay, let's give your job two years. That's all I'll be able to stand."

I hugged him, and Hannah ran to Grannie and shouted, "We're moving to Maneuver!"

WE TOOK THE PLUNGE. I wouldn't have to stick with the mandate of the Forest Service; I could do anything I wanted with whatever grant money I gathered. I could get at the basic questions of relationships in the forest, which had deepened from ideas about connection and communication between trees to a more holistic understanding of forest intelligence.

I taught my first class in the fall of 2002 while still commuting the 380 kilometers each way between Kamloops and Vancouver, while we waited for our new boxy house in the city to close and our log house in the woods to sell. For the first time since Hannah was born, I was alone two nights a week, feeling unmoored. But it was thrilling to have an evening to myself, to go for a walk without a bundle of babies, to read a book without immediately falling asleep, to listen to Jewel on the car stereo without any complaints. On Halloween, we loaded up the truck and moved to our new Vancouver neighborhood, Hannah by now four and Nava two. Hannah adored her lion costume, and I dressed Nava as a calf, and leaving boxes unpacked, we walked down our new block, Hannah yelling, "Trick-or-treat!" for the first time as she ran to doors, toting a pillowcase and copying the throngs of kids. The neighbors of our log house had been too far away, and she'd been too young. Nava nestled in my arms, her head on my shoulder. The kids slept that night in nests of blankets among the boxes in their top-floor bedroom. Don and I watched the shadows of rustling leaves fall across the wall downstairs while listening to footfalls across the sidewalk. Sirens drew closer, and planes descended just over our rooftop, and I wondered what the hell I had gotten us into.

That summer, the policymakers revised their regeneration policy, reducing by half the amount of herbicide spraying in the forests across the province. I was never officially informed, but in time I'd learn that my research drove much of the change.

The first years as a tenure-track associate professor were the most difficult of my life. I was buried in teaching courses, applying for research grants, building a research program, enlisting graduate students, being a journal editor, writing papers. I couldn't let myself fail. Some university mentors told me that a previous female professor had a child, didn't produce enough papers, and was denied tenure. I had brought on a whole new set of worries.

Every day Don and I would wake the kids at seven a.m. and get them ready and take them to daycare and school. I'd work full steam until five p.m., play with them after dinner, work until two a.m. preparing the next day's lectures, collapse in bed, then get up to do it all again. My energy was taxed, I caught too many colds, and on lots of days I felt like a bag of hammers. Don did the rest: picking the girls up from daycare, buying groceries, making dinner, working in between. Being more of a Mr. Mom than he'd imagined. He found it hard to find enough work analyzing data or running models, since the government had reduced funding for forest research. Some of his clients had been from the Forest Service in Kamloops, and he'd lost a few opportunities from not being physically there. He grew irritated about the snarls of the city and spent more time riding his bike on empty roads.

He'd tap on his computer in the mornings, worry about paying the bills, then spend many afternoons with the girls at the Maple Grove swimming pool while I worked on my courses and manuscripts. Interesting work for him came in—once it was modeling how the mountain-pine-beetle infestation was being affected by different forest management practices—but it was not enough. And he was right about raising the girls in the city. We did have to keep a closer eye on them, and we needed to drive them to gymnastics and bike camps rather than just let them play in the forest beside the house. Don took them kite flying and bike riding and to the aquarium and Science World. He bought them slushies and hot dogs. On weekends, we Trail-a-Biked around the city, hung out at the beach, had picnics with our friends, or found a park where they could swing in the rain. But when I succeeded in getting tenure a year beyond our agreed-upon two, our relationship grew strained.

Meanwhile, I was making new discoveries, one question leading to the next. I had grants and students and had won a teaching award. However, as my research program built on success after success toward deciphering the language and intelligence of the forest, my marriage did the opposite, the lines of communication starting to fray and snap. One night after quarreling about Vancouver and Don's unhappiness, I agreed we could move to Nelson. During the semesters, I would stay in faculty housing on weekdays but would commute home to Nelson for the weekends, then back to the city for the next week. Nine hours each way.

It was a difficult compromise, but the dreamy subterranean constellations sprouting in my head as my daughters fell asleep were bearing fruit. My students and I were tracing water and nitrogen and carbon flowing from old Douglas firs to tiny germinants nearby, helping them survive. I was finding proof for my early theories that seedlings deep in the shadows of elderly trees depended on receiving these subsidies through mycorrhizal linkages. I was discovering that the networks in the old-growth forests were far richer and more complex than I'd ever imagined, but in large clear-cuts they were simple and sparse. The larger the clear-cut, it appeared, the more compromised the networks became.

The thought of being in Vancouver in the fall while Hannah and Nava were in Nelson, though, was unfathomable. Little things grated on me—preparing for the field season, dealing with requests to review more manuscripts, coping with year-end reports for granting agencies. One day after work, I rushed to the after-school daycare for the girls and wound through traffic to reach the shop downtown where they'd framed a vellum of maple leaves and pinegrass for me, then charged toward home for dinner. Hannah whined she was hungry, and Nava joined in, and I told them to quiet down, but they screeched higher. "Stop!" I screamed, and I pulled over and hit the brakes. The picture flew into the back of the seats, and the glass shattered. The girls were shocked, and I looked at them in horror to make sure I hadn't hurt them. I pulled them out of their car seats and sat on the side of the

Me, forty-five; Nava, five; Don, forty-eight; and Hannah, seven years old, at our house in Vancouver in the summer of 2005. I had just received tenure as an associate professor at the University of British Columbia.

road and sobbed, my eyes burning like coals. Hannah and Nava cried and wrapped their arms around my neck, and I clung to them. Hannah stopped, and then Nava did. Hannah sniffled and pulled my hair back and said, "It will be okay, Mama."

I took the shattered picture back to the gallery and said that I had dropped it by accident. When they called to say it was fixed, I expected the leaves and grasses to be arranged under new glass—but they'd angled the shards into place, pieced everything back together like a puzzle. I decided I liked it better.

All of it now cracked as intricately as an old face, forever changed.

As we were moving to Nelson, my mind full of worry about living apart while I taught, Dan and I received a grant to create a map of the belowground labyrinth in an old-growth forest. The questions posed were: What is the architecture of this network? Did the pattern help explain the intelligence of nature?

How could we help nurture the young ones without shattering the forest?

12

NINE-HOUR COMMUTE

I pulled into a turnout, slammed down the emergency brake, and grabbed my vest. I ran across the logging road, barely dodging a loaded truck that looked like a giant grasshopper backlit by late morning sun, while singing *"Woohoo"* to warn any bears should they sally along.

Adrenaline pulsed in my ears; I'd found just what I'd been looking for—a hill slope from creek to crest covered with Douglas fir of all ages. The oldest giants looked thirty-five meters in height, branches with plenty of muscle to shower seeds every few years into the shadowy beds packed with needles and humus. The youngsters sprouting from this veil looked like kids in a schoolyard: cohorts of seedlings and saplings flocking and scattering under the watchful gaze of towering teachers. From the road, the tree line seemed as complex as the Manhattan skyline.

I scrambled down the embankment of scree, paused on a knuckle of rock to load my lungs, and leapt across a ditch. A purely Douglas-fir forest perfect for creating a map of a mycorrhizal network. My first graduate student, Brendan, had published his master's work in 2007 showing that a single mycorrhizal fungus, *Rhizopogon,* did indeed coat almost half the root tips of Douglas-fir trees—the other half were colonized, here and elsewhere, by about sixty other fungal species—and it alone formed the major bones of a mycorrhizal skeleton. Young trees were colonized by *Rhizopogon* as were the old, and this was crucial in my quest to understand whether the network helped young Douglas fir establish under the canopy of their elders. Whether the *Rhizopogon* network was key to the continual regeneration of the forest, the ability of the forest to rejuvenate, to sustain itself no matter what. Besides, researchers had already sequenced pivotal portions of the *Rhizopogon* DNA to distinguish one fungal individual—a *genet,* of singular genetic identity, like an individual person—from another, providing a crucial element for making a map of individual fungal strands linking one tree to another. This hadn't been accomplished with other fungal species present in this forest. It was an ideal system to give me a sense, an appreciation of the extent of connectedness. Where, or so I guessed, the young firs might tap into the fungal garden of the old ones. I swished through the grass to the noisy creek and launched off the bank to land two-feet-first on the other side. *"Woohoo,"* I called again, my voice rising above the rush of the water and reverberating more faintly off the escarpment, *"Woohoo, woohoo . . ."*

The trees by the creek were dense and plump, and the ones at the top of the slope looked sparser and smaller. The soil would be drier there, water shedding off the granite knoll like a toboggan sweeping downslope. By comparing the architecture of the network of the dry, upper stand with this moist, lower forest, I could see if the linkages up there, where water was more precious, were denser, more plentiful, more crucial to the establishment of a seedling. Where perhaps the success of the seedlings depended on tapping into the mycelia that were flush with water raised by the taproots of the old trees from the deep granite fissures. Attaching to the mycelial networks of the elders could well be more urgent for seedlings where the soil was parched compared to where it was damp, helping them quench their thirst and gain a foothold.

Striding along the creek, I checked the humus for bear prints. The animal trail along the water's edge had no signs of scat, but I watched for anything unusual beyond the normal shudder of leaves in the blood-dark thickets of red osier dogwood. At the first elder twenty meters in, as I headed up the hill toward the crest, saplings skirting its crown in a configuration like Nava's hula hoop, I pulled out my T-shaped increment corer to check its age, thankful the handle was orange because the leaves of the thimbleberry shrubs were big as dinner plates and could swallow anything that dropped. I fit the bit shoulder-high into a furrow of the tree's chunky bark and cored the tree to the pith, drawing out a small cross section of its striped insides.

Examining the core, my pen dotting each decade, I slowly counted her years. She was 282. I cored another dozen trees around my first one, all different heights and girths, and they ranged in age from five years to the same several centuries as she was. These forests experienced fire every few decades or so, when the summers were dry and there was plenty of fine fuel. When twigs and needles from old trees gathered on the forest floor, and blades from deep grasses senesced and dried, and thickets of new firs started to choke out the watery aspens and birches. With a single spark, patches of the forest would burn, the old trees usually surviving, the understory swept clean. If the fire scorched the floor in tandem with a good cone year, a new bunch of seeds germinated.

I stuffed the tree cores inside colorful straws, sealed the ends with masking tape, and labeled each one so I could double-check the ages and measure the annual radial growth increments under a microscope at my university lab. And I could compare the growth in each year with the corresponding annual rain and temperature records. I ran my thumb across the tip of my trowel to make sure it was sharp and followed a thick root running from the base of my first old tree to where it tapered to the width of a finger, and I sliced open the forest floor in search of rusty-brown truffles, the scabby belowground mushrooms of *Rhizopogon*. The trowel cut through the litter and fermentation layers and slit open the humus to reveal the dense grains of underlying minerals. Where the drizzles of humus and weathered clays came to rest and roots and mycorrhizas foraged for nutrients.

After half an hour, mosquitoes biting my forehead, my knees sore on twigs, I hit a truffle the size of a patisserie chocolate. It was resting smack between the humus layer and mineral horizon, and I scraped away the organic crumbs and found a beard of black fungal strands running from one end of the truffle to the old tree's roots. I followed another pulpy skein in the other direction, and it led me to a cluster of root tips that looked like white translucent pussytoes. The fine, soft brush I'd borrowed from Hannah's paint set was perfect for sweeping them clean. One root tip was especially welcoming, and I gently tugged it, like pulling a stray thread in a hem. A seedling a hand's length away shuddered slightly. I pulled again, harder, and the seedling leaned back in resistance. I looked at my old tree, then at the little seedling in the shadows. The fungus was *linking the old tree and young seedling.*

A shock of nearby boughs shivered, and a yellow butterfly flittered across the meadow. The wind shifted. I looked over at the grasses hemming the fold of trees, blades tingling. My eyes were attuned to edges where consortiums of bears and coyotes and birds linger and banter, but there was no movement.

I tracked another root from the elder and found another truffle, and another. I raised each to my nose and breathed in its musty, earthy smell of spores and mushroom and birth. I traced the black pulpy whiskers from each truffle to the riggings of roots of seedlings of all ages, and saplings too. With each unearthing, the framework unfolded—this old tree was connected to every one of the younger trees regenerated around it. Later another of my graduate students, Kevin, would return to this patch and sequence the DNA of almost every *Rhizopogon* truffle and tree—and find that most of the trees were linked together by the *Rhizopogon* mycelium, and that the biggest, oldest trees were connected to almost all of the younger ones in their neighborhood. One tree was linked to forty-seven others, some of them twenty meters away. One tree bound to the next, and we figured the whole forest was connected—by *Rhizopogon* alone. We published these findings in 2010, followed by further details in two more papers. If we'd been able to map how the other sixty fungal species connected the firs, we surely would have found the weave much thicker, the layers deeper, the stitching even more intricate. Not to mention the arbuscular mycorrhizal fungi adding interstitial components to such a map as they possibly joined the grasses and herbs and shrubs in an independent web. And the ericoid mycorrhizal fungi linking the huckleberries in their own network, and the orchid mycorrhizas with their own too.

A squirrel's midden of seeds was piled against a moist log, so I looked up to the tree crowns for traces of the previous year's cones. Douglas fir makes cones sporadically and in synchrony with the shifts in climate over clusters of years. The seeds are dispersed in summer from yawning cones by wind or gravity, or squirrels or birds, and they germinate in the warm beds of minerals and char and partially decomposed forest floor. Burned mixed seedbeds are especially delicious for germination.

Through the corset of branches, I saw a hawk circle overhead. Solitude is rare in the forest, and I felt slightly uneasy. But the breeze lulled me, and I continued my work, using the finest tip of my Swiss Army knife to excavate a germinant no bigger than a daddy longlegs. I pulled on the collar of the exposed stem, and a radical—one of the tiny primordial roots—slid out of the old-blood humus. It looked like a shard of fine bone china, reminding me of Robyn's shinbone glaring from a ragged cut the time she fell off her tricycle and my father gathered her in his arms. This courageous root was as vulnerable as a growing bone, and it survived by emitting biochemical signals to the fungal network hidden in the earth's mineral grains, its long threads joined to the talons of the giant trees. The mycelium of the old tree branched and signaled in response, coaxing the virgin roots to soften and grow in a herringbone and prepare for the ultimate union with it.

Squatting, I peered at the radical through my hand lens and fumbled to split open the fragile root with my dirt-caked fingernails, to steal a glimpse of the fungal mycelium that might have succeeded in encasing the cortical cells. Finishing the courtship. My nails were so blunt! I twisted around to let the sun pour on my hands, and I scoured the ragged root for signs of tallow between the cells. On invasion, the fungus envelops the root cells, forming a latticework—a Hartig net—the color of beeswax, or seawater, or rose petals. The fungus delivers nutrients, supplied by the vast mycelium of the old trees, to the seedling through this Hartig net. The seedling in return provides the fungus with its tiny but essential sum of photosynthetic carbon.

The roots of these little seedlings had been laid down well before I'd plucked them from their foundation. The old trees, rich in living, had shipped the germinants waterborne parcels of carbon and nitrogen, subsidizing the emerging radicals and cotyledons—primordial leaves—with energy and nitrogen and water. The cost of supplying the germinants was imperceptible to the elders because of their wealth—they had plenty. The trees spoke of patience, of the slow but continuous way old and young share and endure and keep on. Just as

Hannah (right, eight years old) and Kelly Rose (age ten) picking huckleberries in a clear-cut near Nelson, British Columbia, in 2006. The forest is regenerating nicely to spruce and subalpine fir. The height and charring of the stumps indicate the forest was logged in winter and the slash then burned.

the steadiness of my girls steadied me, and I told myself I was strong enough to endure this season of separation. Besides, I'd have a sabbatical in a year, and I could make their lunches again, drumsticks and sliced cucumber and oranges cut into smiles, and I could show them how to build go-carts and plant flowers, and Nava and I could read together more, alternating turns through pages of *Mercy Watson to the Rescue*. But until that magical year, I'd spirit across the mountains each weekend to reabsorb their lives, my motherhood like time-lapse photography.

Once the Hartig net was firmly embedded in the radical of the new sprouts, and the old trees were dispatching sustenance, making up for the paltry rates of photosynthesis by the cotyledons, the fungus could then grow new hyphal threads to explore the soil for water and nutrients. As the miniature crowns of the seedlings spawned new needles, they would feed the mycelium with their own photosynthetic sugars, so the fungus could travel to even more distant pores. Once on solid footing, life running as smoothly as a stock market exchange, the growing root could then support a fungal mantle—a coating—as though donning a jacket of mycelium, from which even more fledgling hyphae could grow into the soil. The thicker the mantle and the greater the number of fungal threads the root could feed, the more extensively the mycelium could laminate the soil minerals, and the more nutrients it could acquire from the grains and transport back to the root in trade. Root begets fungus begets root begets fungus. The partners keeping a positive feedback loop until a tree is made and a cubic foot of soil is packed with a hundred miles of mycelium. A web of life like our own cardiovascular system of arteries, veins, and capillaries. I wound two of my upended seedlings into my hair and started back up the slope.

A cracking sound.

I swept the bear spray out of my holster and fingered the orange safety tab, peering toward a Saskatoon bush. Pulling back a branch, its leaves rustling, I sighed in relief. There was only a stump, its charred bark as black as fur. *Oh, boy,* I thought. *I must be tired from the early morning drive from the coast.*

I continued through the trees, ducking under the crowns of thick-barked elders, striding through grassy gaps sprinkled with seedlings, swimming through thickets of spindly saplings, the data of my graduate students churning through my mind as if in an adding machine. These young trees got their start in the shadow of the old by linking into their vast mycelium and receiving subsidies until they could build enough needles and roots to make it on their own. The Douglas-fir seeds that another of my graduate students, François, had sown around mature trees had a greater survival rate where he'd allowed them to link into old-tree fungal networks than where he'd isolated them in bags with pores allowing only molecules of water to filter through.

The seedlings in this forest were regenerating in the network of the old trees.

Resting on a stump, I took a long drink of water and noticed a cluster of seedlings no bigger than roofing nails. A belowground network could explain why seedlings could survive for years, even decades, in the shadows. These old-growth forests were able to self-regenerate because the parents helped the young get on their own two feet. Eventually, the young ones would take over the tree line and reach out to others requiring a boost.

With the sun straight overhead, I double-checked the time on my BlackBerry. Nelson was still another 476 kilometers away, and to be home by midnight, I needed to leave by four p.m. Jean had urged me to buy this fancy phone—she nicknamed it my BlueBerry—and it was transforming my life, crucial now that I was so many hours on the road. I checked my email, and one of my grant proposals had been rejected. But another one, where we'd submitted looking at clear-cutting of the dry interior Douglas-fir forests and its effects on the integrity of mycorrhizal networks, was approved. *Yes!* I thought. The weeks of parsing words and budgets had paid off. I marveled at this little machine, how the Internet made me feel so connected to the world.

This forest was like the Internet too—the World Wide Web. But instead of computers linked by wires or radio waves, these trees were connected by mycorrhizal fungi. The forest seemed like a system of centers and satellites, where the old trees were the biggest communication hubs and the smaller ones the less-busy nodes, with messages transmitting back and forth through the fungal links. Back in 1997 when my article had been published in *Nature*, the journal had called it the "wood-wide web," and that was turning out to be much more prescient than I'd imagined. All I knew back then was that birch and fir transmitted carbon back and forth through a simple weave of mycorrhizas. This forest, though, was showing me a fuller story. The old and young trees were *hubs and nodes*, interconnected by mycorrhizal fungi in a complex pattern that fueled the regeneration of the entire forest.

Wasps swarmed from a hole beside woody debris. Stung, I ran up the slope, as steep as an escalator, my cruiser vest heavy as a flak jacket, and flopped down at the crest, pressing my water bottle against a welt. The big old trees on this knoll were spaced farther apart, and the saplings were fewer and farther between. Limited by drought. The thimbleberries and huckleberries had disappeared, replaced by the bunched long leaves of pinegrass, the bonnets of silky lupines, and the occasional soapberry shrub. The lupine and soapberry were nitrogen fixers, adding nitrogen to this slow-growing stand. Though the south-facing slope was dry, the plant community was intact, with none of the invasive weeds like the ones creeping along the roadside where I'd parked. This forest was at the northern edge of the arid Great Basin, but to the south it was mainly too dry for trees, and bunchgrasses grew instead in native prairies. These native grasslands were under pressure from exotic weed invasions, and in this case, the mycorrhizal networks were sapping them of life. Knapweeds, spread by cattle, tapped into the mycorrhizas of the grass tillers and stole phosphorus right out of their roots. Instead of the fungi of knapweed helping the grasses thrive, as they had with birch and fir, they were accelerating the decline that had begun with humans herding cattle. They did this possibly by sending the native grasses some poisons or an infection to finish the murder. Or starving them, taking over their energy, degrading the native prairie. Like the invasion of the body snatchers. Or the colonization of the Americas by Europeans.

With my increment borer I cored a handful of ancients on the knoll. The oldest was 302, the youngest 227. The largest, oldest ones were the elders of the forest. Their thick bark was scarred by flames, more pronounced than the trees in the wetter area below, because it was hotter and drier here, a magnet for lightning. This explained the wide range of ages. I checked my phone again. Two o'clock. In an hour, Don would be picking Hannah and Nava up from school.

I scraped the soil with my trowel. Just like the old trees near the creek, those on this crest were decorated with truffles and *tubercules*—clusters of mycorrhizal roots covered in a fungal rind—and golden fungal strands that ran from them like shooting stars. The trees and fungi here too were in an intimate web. Compared to the trees down below, there were even more connections where the soil was drier and the trees more stressed. This made sense! Here on the crest, the trees invested more in mycorrhizal fungi because they needed more from them in return.

I leaned against the oldest tree, at least twenty-five meters tall with branches like the ribs of a whale. Seedlings were germinating in a crescent along the northern dripline of the tree, their needles stretched like spider legs, and I excavated one with my knife. Fungal threads streamed off the end of its roots, and I felt intoxicated, already forgetting the wasp sting. I pressed the seedling and its woolen mycorrhizas between the pages of my notebook so I could look more closely at home. But I already knew; these little seedlings were linked into the network of the old trees, receiving enough water to get them through the driest days of summer. My students and I had already learned that the deep-rooted trees brought water up to the soil surface at night by hydraulic lift and shared it with shallow-rooted plants, helping the archipelago stay whole during prolonged drought.

Without such an attachment, death of a seedling in the hot August days can be nearly immediate, needles turning red and the collars of their stems wounded with burns, leaving not a trace by snowfall. For these young recruits, small resource gains in moments of vulnerability make the difference between life and death, two sides of the dealer's card. But once their roots and mycorrhizas reach the labyrinth of russet pores, where water clings in films to soil particles, they ratchet up their game and grow a foundation. A root system like that, unfettered in its opportunity, was far more resilient than the chunky pistons grown in Styrofoam tubes in the nursery, where the seedlings intended for plantations were so stuffed with water and nutrients they couldn't—didn't need to—sprout adequate roots to partner with fungi to connect with the soil. Their thick needles needed streams of water under the hot August sun, but their roots continued to grow as though imprisoned, unable to reach the old trees for help when the soil cracked in the dry clear-cuts.

I walked from the northern crescent of seedlings back to the old tree, the ground directly underneath its canopy bare even of grass. Not a seedling grew here. Its crown was so dense that it intercepted most of the precipitation and sun, and its roots were so thick that they took up most of the nutrients and water. But François would later find there was a sweet spot, a donut, at the dripline, the fringe of the crown, where the water dripped off the outermost needles, where some seedlings flourished. Not too close to be starved by the needs of the old tree, and not too far away for the grasses in the intervening meadows to rob them of what they required.

I ducked under the opposite edge of the old tree's crown—facing south, where the sun beat down—and gazed down the slope rolling into scree. It was so hot and dry on this side that not even a network could save a seedling from burning up. In the extremes—such as a desert—even the fungus could fail to bring life to a tree. An old log lay on the angle of repose, poised to roll over broken stones, and chunks of heartwood were newly exposed, beetles and ants flowing in lines with white fungi in their clutches. Claw marks. Bear, I thought, from at least a few days before. Douglas-fir seedlings cascaded off the north side of the log, where there was a sliver of shade along its length, and they spilled onto the forest floor. The scrap of advantage from the shade meant a little less water lost, a slightly thicker film coating the soil pores, the difference between survival or not. I wondered if the white fans of mycelium were linked to the old tree and helped keep the wood moist. These seedlings were alive, I figured, only because the fungi were importing water from somewhere.

Skin burning, I returned to the shade and checked the wasp sting. I should show my girls how to make a poultice of baking soda. I sat and leaned back against the old tree nurturing that crescent of seedlings through the mycorrhizal network, the needles of the young quivering in the afternoon air.

The old trees were the mothers of the forest.

The hubs were *Mother Trees.*

Well, mother *and* father trees, since each Douglas-fir tree has male pollen cones and female seed cones.

But . . . it felt like mothering to me. With the elders tending to the young. Yes, that's it. *Mother Trees. Mother Trees connect the forest.*

This Mother Tree was the central hub that the saplings and seedlings nested around, with threads of different fungal species, of different colors and weights, linking them, layer upon layer, in a strong, complex web. I pulled out a pencil and notebook. I made a map: Mother Trees, saplings, seedlings. Lines sketched between them. Emerging from my drawing was a pattern like a neural network, like the neurons in our brains, with some nodes more highly linked than others.

Holy smokes.

If the mycorrhizal network is a facsimile of a neural network, the molecules moving among trees were like neurotransmitters. The signals between the trees could be as sharp as the electrochemical impulses between neurons, the brain chemistry that allows us to think and communicate. Is it possible that the trees are as perceptive of their neighbors as we are of our own thoughts and moods? Even more, are the social interactions between trees as influential on their shared reality as that of two people engaged in conversation? Can trees discern as quickly as we can? Can they continuously gauge, adjust, and regulate based on their signals and interactions, just as we do? Just from the inflection of the way Don says "Suze," and from his brief glances, I comprehend his meaning. Maybe trees relate to one another as delicately, with such attunement. Signaling as precisely as the neurons in our brains do, to make sense of the world. I scribbled quick calculations based on our isotope work. It occurred to me that the amount of carbon transferred relative to nitrogen was strikingly similar to their respective quantities in molecules of an amino acid called glutamate. We hadn't exactly sought in our experiments to trace glutamate's carbon-nitrogen movements, but other researchers had verified that the amino acid itself did move through mycorrhizal networks.

I did a fast search on my BlackBerry. Glutamate was the most abundant neurotransmitter in the human brain, and it set the stage for other neurotransmitters to develop. It was even more abundant than serotonin, whose carbon-to-nitrogen ratio was only slightly greater.

The hawk arced around the hill next to mine, now joined by two others, casting their shadows across the scree-pocked forest. How similar could the mycorrhizal network *really* be to a neural network? Sure, the pattern of the network and the molecules transmitting from node to node through the links might be similar. But what about the existence of the synapse; isn't that crucial to signaling in a neural network? This could also be important in a tree detecting whether its neighbors were stressed or healthy. Just as neurotransmitters pass signals across the synaptic cleft from one neuron to another in our brains, signals might also diffuse across a synapse between interfacing fungal and plant membranes in a mycorrhiza.

Could information be transmitted across synapses in mycorrhizal networks, the same way it happens in our brains? Amino acids, water, hormones, defense signals, allelochemicals (poisons), and other metabolites were already known to cross the synapse between the fungal and plant membranes. Any molecules arriving by way of the mycorrhizal network from another tree might also be transmitted through the synapse.

Maybe I was onto something: *both neural networks and mycorrhizal networks transmit information molecules across synapses.* Molecules move not just through the cross walls of adjacent plant cells and the end pores of back-to-back fungal cells, but also across synapses at the apices of different plant roots, or different mycorrhizas. Chemicals are released into these synapses, and the information must then be transported along an electrochemical source-sink gradient from fungal-root tip to fungal-root tip, similar to the workings of a nervous system. The same basic processes, it seemed to me, were occurring in the mycorrhizal fungal network as in our neural networks. Giving us that flash of brilliance when we solve a problem, or make an important decision, or align our relationships. Maybe from both networks emerge connection, communication, and cohesion.

It was already accepted widely that plants use their neural-like physiology to perceive their environment. Their leaves, stems, and roots sense and comprehend their surroundings, then alter their physiology—their growth, ability to forage for nutrients, photosynthetic rates, and closure rates of stomata for saving water. The fungal hyphae, too, perceive their environment and alter their architecture and physiology. Like parents and children, my girls and Don and me, adapting to change, aligning to learn new things, figuring out how to endure. I'd be home tonight. *Mothering.*

The Latin verb *intelligere* means to comprehend or perceive.
Intelligence.

The mycorrhizal networks could have the signature of intelligence.

At the hub of the neural network in the forest were the Mother Trees, as central to the lives of the smaller trees as I was to Hannah and Nava's well-being.

It was getting on, so I got up, sorry to leave the bark warm against my back. But I was breathless with elation, high on my thoughts, and I felt a kinship with the Mother Trees, grateful for accepting me and giving me these insights. I walked to the top of the knoll, remembering a small route out to the main haul road, and I followed a deer trail heading roughly in its direction. The thick fungal strands of the tough *Rhizopogon* truffles and the fine mycelial fans of the fragile *Wilcox-*

Douglas fir Mother Tree

ina mushrooms, and the hundreds of other fungal species in this old forest, had unique structures and abilities to acquire, transport, and transmit. Their long threads reached for treasure, their tendril fingers wrapped around the prizes. The information chemicals must transmit through these fungal highways along the various routes. Following the source-sink gradients between rich and poor.

My little trail joined another, like a frayed thread joining a rope. I knew the networks were complex, with thick cords like freeways amid a gauze of fine hyphae that behaved like secondary routes. The thick cords themselves consisted of many simple hyphae that had twined together, forming an outer rind around a space. Information chemicals could travel through these cords like water through a pipeline.

The main trail widened, and after a few more curves, the small road would lie ahead. The thick pipelines of fungal species like *Rhizopogon* were designed for long-distance communication, and the fine mycelial fans of fungal species like *Wilcoxina* must be adept at rapid response. Able to transmit chemicals swiftly to trigger fast growth and change. When Grannie Winnie was diagnosed with Alzheimer's, I'd read about what makes our brains either plastic or rigid. Maybe the long-distance *Rhizopogon* were analogous to the strong links in our brains arising from repetition, pruning, and regression, giving us long-term memory. Maybe the finer *Wilcoxina* hyphae, which grew faster and more abundantly, helped the mycorrhizal networks adapt to new opportunities, not unlike our own rapid, flexible responses to new situations, which Grannie was losing.

Grannie Winnie still had long-term memory. She knew she had to put on clothes; she just couldn't remember how many shirts to wear when it got hot, or whether to clip her bra in front or behind. Just as *Rhizopogon* strands deal with long-distance transport of solutions, Grannie's memory about wearing clothes came from lifelong brain pathways. But her ability to adjust quickly, and her short-term memory, were dwindling with the loss of new synapses, as if she were losing connections analogous to the ones created by the mycelial *Wilcoxina* fans for trees.

The thick complex strands running out from the Mother Trees must be capable of efficient, high-volume transfer to the regenerating seedlings. The finer spreading mycelia must help the new germinants modify to accommodate pressing, rapid needs, such as how to find a new pool of water on a particularly hot day. Pulsing, active, adaptive in providing for the growing plants, like fluid intelligence.

The new grant would eventually show us that the complex mycorrhizal network unraveled into chaos with clear-cutting. With the

Mother Trees gone, a forest would lose its gravitas. But within a few years, as seedlings grew into saplings, the new forest would slowly reorganize into another network. Without the pull of the Mother Trees, though, the new forest network might never be the same. Especially with widespread clear-cutting and climate change. The carbon in the trees, and the other half in the soil and mycelium and roots, might vaporize into thin air. Compounding climate change. Then what?

Wasn't this the most important question of our lives?

I reached a colossal tree, a rampart, her branches thick right to the ground and as big as trees themselves. Her large size and old age were magnificent compared to her neighbors. She looked like the mother of all Mother Trees. What foresters call a "wolf tree"—far older, bigger, and with a much wider crown than the others, a lone survivor of previous calamities. She had lived through centuries of ground fires that others had—at one time or another—succumbed to. I waded through squalls of seedlings to get to the fringe of her crown and picked up a cone perhaps clipped by a squirrel, its bracts dusted in white spores. Her life had started when the Secwepemc people cared for this land, long before the Europeans came, when the Native people regularly lit fires to create habitat for game, or to stimulate growth of valuable native plants, or to clear routes for trading with neighboring nations, and they'd kept the fuels low so the flames were never intense enough to have burned off her thick bark completely. I was sure if I cored her, her rings would be calloused with char every twenty years or so, like the stripes of a zebra. I was struck by her endurance, her rhythm that spanned centuries. It was a matter of survival, not a choice, not an indulgence. Light glanced off her bark, incandescent, the sun dropping.

Brilliance.

I returned to the trail, telling myself to publish my thoughts on Mother Trees as soon as I could, and I rounded the last corner before the road.

On the edge of the path, only two meters away, were two cubs the size of teddy bears, staring through the purple larkspurs and pink lady-slipper orchids. One was brown, the other black, and they looked

Working, age forty-seven, while on the road in our VW van, 2007

at me politely. Behind them wearing a black fur coat was their mama. She growled, and they darted into the huckleberries and birches, leaving me stunned. Alone, untouched.

I hurried onto the little road and raced to the main haulway, wondering if they'd been with me all day.

HEADING OVER the Monashee Mountains, I crawled around hairpin turns, dusk falling.

The taillights in front of me swerved.

Legs. Long legs, as tall as my truck, attached to the torso of a moose.

My reactions were hampered from fatigue, but I yanked the steering wheel to my left, then slowed. As I passed the moose, I looked through the windshield, straight into her eyes before they—she—slid into darkness. Old eyes that had seen right through me. They knew I couldn't sustain this pace.

I pulled into our Nelson driveway at two a.m., so tired I felt hit by a van. I snuck into Hannah's room, and she stirred when I kissed her forehead. I crawled under the covers with Nava. Her bed, moved from Vancouver, was barely big enough for both of us. I held her hand between both of mine, and I could have sworn her fingers were longer since last weekend. She squeezed my hand.

My sabbatical in 2008 brought the relief I'd imagined, and I published two papers on the Mother Tree concept. But I returned to work the following fall, and the endless nine-hour commutes started again. The girls attended school and danced, and Don watched out for them and skied and got an occasional computer-modeling job, but I grew increasingly fatigued, and Don and I quarreled more.

My lab was busy, and I chased after grants and wrote more articles. I continued to teach and work on the free-to-grow problem, and I published three journal articles in 2010 showing that free-to-grow lodgepole-pine plantations were in jeopardy with the warming climate. Jean helped me collect the data and Don analyzed it, and we'd found that more than half the pines in the province were dying from insects and disease and other maladies like drought stress. More than one-quarter of the plantations would be considered understocked.

On my annual drive home from a fall research camp at the end of August 2010, not long after I'd presented the pine work at a provincial conference, I stopped at a gas station and checked my iPhone. There was a message from the policy guys, arguing that we'd used obsolete methods to measure infection by western gall rust, one of the fifty damaging agents. Infections on branches were now considered deadly only if they occurred two centimeters from the main stem, not four. How they'd suddenly figured out that infections at four centimeters weren't a problem but they were at two centimeters was odd; their discovery had been made the moment we'd published our papers. Another independent study, though, had verified that the majority of pine plantations were in poor health. But what got me most was an email from a long-respected government statistician, someone I admired and who'd approved our sampling approach, suggesting our design had not been replicated a sufficient number of times.

Crisscrossing the mountains between Vancouver and Nelson, watching the beetle-killed forests turn into a mange of clear-cuts, my anger with forestry practices grew. I coauthored a *Vancouver Sun* opinion editorial with Dr. Kathy Lewis, a colleague at the University of Northern British Columbia, that we titled "New Policies Needed to

Save Our Forests." We highlighted the sea of clear-cuts, citing how they were "reducing landscape complexity and affecting broad-scale ecological processes such as hydrology, carbon fluxes, and species migrations." We wrote about young, simplified forests planted solely with single species that were declining due to insect, disease, and abiotic damage, and said this would worsen with climate change. Deep cuts to funding forest science had greatly reduced British Columbia's capacity to assess the true state of our forests and respond appropriately. We ended with a rallying cry for changes in policy to increase the resilience of British Columbia's environment and economy. We followed this article with another one that suggested how to fix the problems.

The morning that first op-ed appeared, I was home, pacing our living room, imagining the vitriol from the capital. I was tired, but I was on fire. Over the course of the day, a hundred foresters wrote responses to the newspaper in agreement, one saying, "Thank you, Kathy and Suzanne, for an excellent and accurate portrayal of British Columbia's dirty little secrets." I petitioned the Ministry of Forests to restore research funding across the province, gathering signatures from dozens of colleagues. "Brava," wrote a professor emeritus from the University of British Columbia, but few professors signed.

At home on the weekends, I was unable to sleep. One night driving over the passes, I hit a deer. On another, my alternator packed it in at minus twenty Celsius, and I coasted down the mountains, barely making it to a garage.

Late one Sunday night driving back to work, dark circles instead of eyes reflecting in the rearview mirror, I knew I couldn't do this any longer. Don had come to the end of his rope too. I'd drowned in the stress of the commute, and he had grown even more frustrated that I wouldn't quit. "We love you with all our hearts, but your dad and I have decided to separate," I told Hannah, fourteen, and Nava, twelve, in the living room on July 20, 2012. Don was pale, and I crouched lower, yearning for wholeness, wanting to protect Hannah, who sat stunned, and Nava, completely lost, staring at her sister.

Don managed to sit straighter and come up with, "It'll be fun. You'll each have another bedroom!" Hannah brightened and asked if she could have a double bed. Nava looked at Hannah and bounced once or twice on the couch.

With the help of Mum and a dose of good luck that a little century-old house not far from Don's had just come up for sale, the girls and I were able to move soon afterward. We painted Nava's room robin's-egg blue and Hannah's creamy yellow and the tiny balcony off Nava's upstairs bedroom lime green, and we sat there in the evening to gaze at the mountain across the lake. I held the girls close and breathed in their childhood scent, and sometimes we fell asleep out where the mountain air swept away the day. I wished I could have protected them from the breakup, but I knew in the long run they'd rather have a healthy mother and happy father. By midsummer, with temperatures soaring and the woods brittle from drought, fires were burning across the province, and smoke hung in the valley.

13

CORE SAMPLING

"Plenty of time to get to the top and back by dark," Mary said, starting up the cinder trail to Tam McArthur Rim in Oregon.

The afternoon sun was high. I was still becoming attuned to "Mary time," which meant leisurely starts after creamy coffee and a sifting through maps to plan a hike. I was used to rushing—hauling loads of kids, food, and packs into the car for even short treks—but today we'd had a slow start, picking tomatoes and cucumbers from her garden for our lunch. She knew each contour of the trail, how long it took to reach her favorite vista, and how much time we could spend tending her squashes and beans.

"We're JIT," she said, tilting a smile at me as we reached the trailhead at two p.m. JIT meant "just in time," a cherished ingredient of our adventures. She strode like she belonged, as comfortable as the old pines among the cinders, with her frayed bootlaces battened down, tattered fanny pack cinched, and straw hat ties knotted under her chin. She wasn't the least bit bothered by the younger hikers already coming down and shedding space-age packs. The rim of the basaltic plateau was a thousand feet above, a bald eagle soaring over the weather-beaten trees. How lovely—perfect—to spend the evening alone with her on this trail. I wanted more of Mary time. I gently pushed her shoulder and said, "We'll get to the ledge by sunset."

Mary was our neighbor when Don and I were doing our doctorate studies in Corvallis. I'd stayed with her for a few days in late August while I presented a paper on mycorrhizal connections at a conference. Our evening conversations had drifted over backpacking and canoeing routes, books we'd read, movies we'd seen, how Nava was already in eighth grade and Hannah in tenth, and how she hadn't seen them since kindergarten, and we'd visited the whitebark pines in the Oregon Cascades. "Maybe you can show me a Mother Tree," she'd said, having listened to me jabber about my recent discovery. Mary was a veteran hiker, having grown up in the Sierras of California, and she'd settled in Corvallis to work as an R & D physical chemist after a postdoc in Australia. I told her she could help me figure out which chemicals were moving through the network. She'd been living by herself all these years, focusing on her job of developing inks for inkjet printers and recovering from the car accident that had taken her friend and injured another and left her badly hurt too.

"What are these globs?" Mary asked, pointing to drops of yellow pitch on the bark of the dead lodgepole pines lining the trail.

"Pitch tubes from the mountain pine beetle," I said, catching my breath in the thin 6,600-foot air. I could barely keep pace, even though her right leg had been screwed together with plates and was a good inch shorter than her left. I pinched off a piece of the resin, hard like old chewing gum, and put it in her hand. "Is that why the pine is dead?" she asked, wisps of blond hair escaping her ponytail, sunglasses secured with a lanyard. I explained that the pine tried to pitch out— eliminate—the beetle when it burrowed into its bark, but the ultimate cause of death was the blue-stain fungus carried into the wood on the bug's legs. The pathogen spread through the xylem, plugging the cells and cutting off the water coming up from the soil.

"The tree died of thirst," I said.

"Geez, dying isn't that straightforward for a tree," she said, offering me a drink from her water bottle before taking a swig herself. "I would never have guessed that."

We took in all the dead trees, far as we could see, some whose needles were red, others still green. Lupines remained bright purple among the gray stems, and the grouseberry shrubs shone, taking advantage of the unused sunshine and water, their fuchsia berries sweet as raspberry jam. "The beetles kill the old pines, and then fire melts the resin in the cones to release the seeds. That's why the young lodgepoles grow in dense thickets after a burn." I put some berries, not much bigger than raindrops, on her palm and pointed to a clump of juvenile pines, saying that these forests used to be patchy—a mosaic of differently aged stands, some old, but most too young to support an infestation. "Things are different now," I said, explaining that suppression of fire had allowed many trees to reach such an old age and large size that their phloem was thick enough to support a teeming brood of larvae. The beetle outbreak had started in northwest British Columbia and spread south to Oregon, with more than 40 million hectares now dead or dying across North America.

Even though the beetle and fungus had coevolved with the pines, the past few decades of fire suppression had created a vast landscape of aging pines ripe for an epic infestation. With winter temperatures no longer dropping to minus thirty degrees Celsius for long enough periods to kill the larvae feeding in the phloem, the finely tuned symbiosis among the species had ruptured. We were in an outbreak of such massive proportion that people in its midst were reeling.

"Are all these trees going to die?" she asked, starting back up the trail, rusty dust powdering her calves, bare arms muscled from bringing in the winter wood, gait long adjusted to the realignment of her bones.

"Some will live, but most will die," I replied. The pines produced an array of defense compounds—*monoterpenes*—to inhibit the beetles. I loved that she was worried for these trees too. Sweeping her hand against a dead trunk, she grabbed a handful of red needles for my inspection. "This outbreak is so intense most trees can't fend off the bugs. They've even detected the swarms with satellites," I said.

She pointed at a small patch of pines with needles deep jade and suggested the future might not be wholly bleak. I agreed, a bit sheepish. The sweep of dead trees throughout the West was disturbing to witness. Some individual pines could increase their defense through a greater production of the monoterpenes, but even so, not many had survived this outbreak. Subalpine firs under the dead pines had put on new growth, though their needles and buds had been chewed by the western spruce budworm, another insect infesting the conifer forests of the West. Still, in spite of the worms burrowing in the buds of the firs and the beetles in the pines, the forest here was anything but dead. Many saplings were in good health, and plants were spreading into the gaps where the dead pines had fallen over. "Survivors should produce new generations better adapted to pitch out the beetles," I said. I needed to take a longer view instead of being so obsessed with the dying trees. Mary took my arm and said, "You'll see, Suzie, it'll get better eventually." *She's right,* I thought. Still, things had gone badly off the rails—pines in valleys from the Yukon to California were utterly dead.

"It's even possible the firs and pines can warn each other about infestations," I said as we continued along the trail. I explained how Dr. Yuan Yuan Song, a scientist from China, had been working with me to see if firs infected with budworms might warn neighboring pines to prepare themselves. Her query had arrived out of the blue, asking if she could come for a five-month postdoc to test whether the warning system she'd detected among tomato plants in the lab also occurred among coniferous trees in forests. Yuan Yuan had already found that tomato plants communicated their stresses to other nearby tomatoes, and we were both curious if similar signals might occur between trees.

A whiskey jack flew in front of Mary and chuckled *quee-oo.*

We were on the plateau in under an hour, the subalpine firs dwindling as we wound through meadows of beargrass and volcanic rock. Mary put shiny chunks of obsidian and feathery pieces of pumice into her pack, and she slipped a few into mine. "Nava will like this one," she said, shining it on the hem of her T-shirt. We reached the rim and followed the trail skirting its edge, columns of basalt stretching down cliffs. Clumps of thousand-year-old whitebark pines followed the contour of the escarpment, forming the tree line.

I showed her the five-needle bundles growing on the branches of the whitebark pines, distinguishing them from the fascicles of only two for lodgepole pine.

Whitebark depends on Clark's nutcrackers to disperse its seeds, whereas lodgepole needs fire to open its cones. As if on cue, a gray-and-black bird swooped from a tree with a cone in its beak and flicked over the lava flow, probably to cache it in a favorite hollow among the rocks, which explained why the whitebarks usually grew in clusters. These two species were in a mutual relationship, the birds dispersing the seeds to fecund soils in return for a stockpile of nutritious meals, having coevolved in the harsh alpine environment, both creatures' genes rigorously shaped through recombination and mutation, adapting bit by bit to glacially slow changes.

"Are these whitebark pines Mother Trees?" Mary asked, skirting a patch of three wrinkled ones with branches stretching in the direction of the wind. Last night we'd watched *Mother Trees Connect the Forest*, a short documentary I'd made with a graduate student along with a filmmaker who was also an adjunct professor at the university, and she was trying to compare these subalpine trees to those in a rain forest. I pointed to the tallest one of the bunch and said that Mother Trees are the biggest, oldest ones. I grabbed her hand to duck under its crown, to see if the roots were wrapping around those of its neighbors. Mary gestured toward a spate of seedlings on the edge of the canopy. This copse of trees, their thick roots woven in sprawling runners, were sure to be united by a mycorrhizal network.

We reached Mary's favorite ledge as the sun dropped in the west, the eight-thousand-foot rim throwing shadows onto the red and green forest below. I knew then that the next step in my pursuit was to find out if trees warned one another of disease or peril, whether the dying species would persist or a different one would take over territory. Mary pulled out the wraps she'd made with the tomatoes and cucumbers, and I uncorked our wine as the string of ancient volcanoes—the Three Sisters to the south and Jefferson, Washington, and Adams to the north—turned from yellow to pink. They stood like monuments, their peaks dwarfing the adjacent slopes. They were different from the Rockies back home, where the peaks were piled closely, metamorphic rock and sedimentary layers tilted together, one arête crowded next to another. Mary's face absorbed the last rays of sun, both of us enjoying this freedom, the company of each other. I had that old sense of falling, softly and deeply, like snow settling on the mountains.

The next morning, she picked earthy-sweet blueberries and mixed them with blackberries, and we ate under the shade of her quince tree. She read me an excerpt from Ken Kesey's *Sometimes a Great Notion* and invited me to canoe down the Willamette River in the fall. I didn't want to leave; every cell in my body stirred. I lingered until there was barely enough time to make it by midnight to the site where I would teach my field course the next day, eleven hundred kilometers to the

Examining western hemlock roots in Robyn and Bill's backyard in Nelson, British Columbia, in 2012. Western hemlock forms shallow root systems, which help the trees forage for scarce nutrients from the young glacial moraine soils. Like many people in British Columbia, Robyn and Bill live on the edge of the forest. They have cut the smaller trees from the understory to reduce the fuel load and the probability that an errant fire will torch the crowns of the trees and burn their house down. Fire risk has increased rapidly in the small towns of British Columbia due to climate change.

north. *Mary time. Was I in love?* A hundred kilometers past the Canadian border, the cold of the fall snapping at the trees, I stopped at a phone booth and called her. Snow shocked my bare arms still warm from the Oregon sun. I said I would come canoeing once I was back at the university in September.

The line hummed with the sound that the deepest quiet makes.

"I can't wait," she said.

A WEEK LATER, driving back to Nelson to help Hannah and Nava get ready for school, I passed a hundred miles of gray trees dead from the mountain pine beetle. Along the route, west of Kamloops, a ponderosa pine stood alone, her weary red crown hunched. I wanted to see how old she was when she died, and if anything was regenerating to replace her. As I walked to this Mother Tree, her withered needles crackled under my feet. No nuthatches sang *qui-qui-qui-qui* from her outstretched arms.

I inserted the bit of my increment corer into her orange-brown bark, but the coil wouldn't enter the dry cork. Flakes shed like random puzzle pieces, with bleached wood beneath the cambium. Desiccated cones hung from her fingertips, scales open, seeds flung, her last gasp. From the look of things, she'd been dead at least a year. At my feet was a nest of fine bones and broken eggshells that must have fallen from her limbs. The soil was dry and deeply cracked. Death had passed down the chain, taking the squirrels and fungi too. Across the Thompson River valley, the air was thick with the smoke of a wildfire, the river a charcoal shade instead of blue. All the ponderosas sandwiched between the grassland along the valley bottom and the Douglas firs in the montane were dead. The firs had a tint of blood red, having been chewed by the western spruce budworm. The death in the forest reminded me of what Mary and I had seen on Tam McArthur Rim, but she would have reminded me that there was still life here.

My BlackBerry read three p.m., seven hours before I'd arrive home. I checked for seedlings around the bony skirt of the dead Mother Tree and found a few two-year-olds huddled in a crevice. These siblings were all that remained to carry the Mother Tree's genes. I knelt to examine them, grasshoppers jumping from under the drooping awns of cheatgrass—a native plant run amok in the barren soil. If whitebark pine seedlings could live in the cold soil of the subalpine, surely these ponderosa seedlings would survive down here. At this age, they should be firmly clasping the earth, but fungi and bacteria were no longer gluing the sand and silt grains into clumps, giving the soil structure to hold water. I pushed the metal probes of my new soil-moisture sensor—a great advancement since the neutron probe—into the loose dirt to measure the soil water content. Only 10 percent registered, barely enough. It seemed incredible that these seedlings were surviving. Perhaps their mycorrhizas were drawing scant water from the dry grains. The broken bones of the Mother Tree still cast some shade, and I wondered if she'd been around long enough to help. I'd read that dying grasses shuttled phosphorus and nitrogen to their off-spring through arbuscular mycorrhizal networks and wondered if this Mother Tree did the same as she died. Sending them her last drops of water, along with some nutrients and food.

The trees had perished so swiftly, and the beetles had spread so rapidly and the summers warmed so quickly, that it seemed nature had no time to figure this out, to keep up with the changes. *Quelle tristesse.* Even if these seedlings survived their childhood, they would likely be maladapted before they were saplings, susceptible to infections and infestations and likely doomed by the changes the climate scientists were predicting. The ponderosa-pine woodlands were turning to grasslands, while the Douglas-fir forest was being overtaken by the ponderosas.

Was this the best the forest could hope for? More likely, the cheatgrass—along with the spotted knapweed and common burdock—would be more successful than the trees at filling the parched earth, at least down here in the valley. These plants, with their prolific seed production and rapid growth rates, could easily invade a forest weakened by fire suppression and extreme climate. These trees seemed sacrificed for the sake of human convenience. Ironically, the very weeds and insects that were killing the forests might be the ones with the genes to persist as the temperatures rose and the rains changed.

The sun draped crimson over the chewed crowns of the distant Douglas firs. The ponderosas that mingled among them, though, glimmered like emeralds, still alive at higher elevation, survivors of the beetle outbreak. I guessed it was less stressful for the pines upslope where the rainfall was more plentiful. But unlike the pines in this *ecotone*—this transition area between the lower and upper forest communities—the firs were feeling the drought, their taproots not reaching as deeply into the parent material, reducing their ability to resist infestation by their own coevolved herbivores. Maybe this was why they seemed so heavily defoliated by the budworm.

Were the ponderosas doing well because of their deep taproots and more plentiful rainfall up there, or was it because of links to their Douglas-fir neighbors? My old doctoral supervisor, Dave Perry, had already found that the two species were likely joined in a mycorrhizal network in the forests of Oregon, and he'd thought that Douglas fir shared enough nutrients to affect the growth rate of ponderosa pine. I figured that was likely going on here too.

Integration of the two species in a mycorrhizal web could well provide more than just an avenue for swapping of resources. If the Douglas firs dying in the drought were making way for pines better adapted to the warming temperature, *were they still connected and communicating with the pines, even as they passed?* Could the firs *warn* the pines that there were stresses in the new areas? Maybe they could send the pines information about sickness.

My colleague Yuan Yuan's tomato plant had not only transmitted warning signals through the interlinking arbuscular network to its neighbor about a pathogen infection, but the neighbor upregulated its defense genes in response. Even more, the neighbor's genes got to work and produced an abundance of defense enzymes. These enzymes must have subdued the pathogen, because when the fungus was applied to the eavesdropping tomato, it incredibly did not cause disease. Yuan Yuan had come to help me ask the same of ailing firs, to see if the pines stood a better chance in their new environment because the firs signaled the nature of their plight.

I picked up two pieces of the Mother Tree's puzzle bark, one for Hannah and one for Nava, and put them on my dashboard for luck. I sped over the Monashee Pass, the start of night blurring the outlines of the road before my eyes adjusted to the headlights. By the time the ferry landed me on the other side of the Arrow Lakes, I was dead tired. "Watch for deer at dusk," Grannie Winnie always warned, bringing a familiar anxiety. I touched the lump I had discovered recently in my breast and reminded myself to go back to the doctor soon. I was sure she was right; it was nothing. My last mammogram had given me an all clear.

"PASS THE EIGHTEEN-GAUGE," the oncologist said to the nurse, pointing to a fine, short needle on the tray.

I lay facedown on the elevated operating table, my left breast hanging through a moon-sized hole, making it accessible from below. The smell of antiseptic and body odor was overpowering in the tiny biopsy room. I wanted to flee to the sweet shade of a sprawling Mother Tree—not caring if she was dead or alive. In front of me was a screen showing the white spider in my breast. I repeated the mantra Jean taught me: *Every little thing is gonna be all right.* I was a forest dweller, backpacker, backcountry skier, organic-food eater, nonsmoker, mother of two breastfed children. Mary squeezed my hand and whispered, "You're going to be fine."

The needle gun went *POP,* and pain seared my breast.

"Hmm. Pass the sixteen-gauge," the doctor said.

The nurse picked up a bigger one. The needles were in a row, from thin and short to thick and long, and reminded me of the ones I'd used to inject $^{13}C-CO_2$ into the bags Dan and I had sealed over my seedlings. They each had a cutting sheath like an Oakfield core for collecting soil samples, extra sharp to slice through roots. Mary read the screen and watched the needles, leaning a little toward the wall. Though brave enough to hike up Tam McArthur Rim at all hours, she crumbled when someone was in pain. I'd never forgotten that letter she'd sent when Kelly died, saying how sorry she was, that she knew how much I was suffering, that sometimes it gets worse before it gets better. Her kindness had made me feel less alone with my grief.

"This lump is hard as a rock. I can't get this needle in either." Tension rose in the doctor's voice. "Let's try the fourteen-gauge."

The word "disease" popped into my mind. A disorder of the body. *Every little thing is gonna be all right.*

"Okay, that's one sample, four to go." Sweat glistened on the doc's forehead, and his breath was stale with coffee.

Four more? That didn't sound good. The nurse shuffled the instruments. Mary's fingers were getting slippery, but I hung on to them as if I were tumbling off a cliff. Think of those whitebark pines we'd hiked through, the Mother Trees, the survivors of beetles and rusts, their groves of offspring where snow lay long into the summer.

Needles passed expertly from one pair of hands to another, few words spoken.

"I don't know where this ends," he said grimly.

The blood drained out of my head. What the hell did that mean? Mary dropped my hand, and the nurse rushed to help her into a chair. The doctor abruptly shed his surgical gloves, said I'd get the results in a week, and walked out. The nurse murmured something soothing, and Mary fumbled to help me button my shirt. She was always steady, but her fingers shook.

When we sank into my car in the alley behind the clinic, I panicked. What should I do? Call Hannah and Nava? *Oh God.* What if I had cancer?

"We don't want to alarm the girls," said Mary. She grabbed my wrist and told me to breathe slowly through my nose. "And we have to wait for the biopsy results before we know anything for sure."

I turned the key in the ignition, but she stopped me and said, "No, let's wait until you're calm." Mary time again. I wrapped my arms around the steering wheel and leaned onto it while she kept her hand on my back. I would have torn out of the parking lot, trying to run away from this, making it worse.

In my apartment on campus, I cried and clung. Kids were shouting in the playground. My plants on the sill were straining toward the light, and I got up, as though on autopilot, to give them cups of water. I called Mum and Robyn, and Mum's cousin Barbara, a nurse who had survived breast cancer. She promised to keep tabs on this. Jean, not able to hide her worry, said, "You'll be okay, HH," HH standing for "Homer Hog," a nickname she'd given me in our university days because I loved digging around in soils like a groundhog. I softened at her crooning my nickname. I floated through the apartment, in some other dimension, while Mary told me I had to be hungry and made chicken mole, banging pots and raiding my cupboard for chocolate and canned chilis.

She was so right. I leaned against her; I was ravenous.

"THIS LITTLE LUMP OF MINE, I know it's benign," Mary and I sang to the tune of the children's song "This Little Light of Mine." Robyn had urged me to sing it whenever I started to feel anxious. Mary and I were climbing a steep path around refrigerator-sized stones and mountain hemlocks with pistol-butted trunks, bent from the downward creep of snow. We had stuck with our weekend plans to hike the Sigurd Peak trail at the confluence of the Squamish and Ashlu Rivers near Vancouver, a damn sight better thing to do than fretting at home. Don and I had agreed not to tell the girls, given the low chance of cancer. What they didn't know couldn't hurt them.

The switchbacks were a good distraction, my steps short and deliberate, the song on repeat, my worry coming in fits and starts. The hemlocks didn't seem the least bit concerned, and I appreciated their calm demeanor. They were built for duty, clinging like mountain goats to rocks, tossing cones like pennies, unafraid of the worst. On the mountaintop, we caught glimpses of glacial ice flowing from peaks, helpless to resist climate change. I wanted to keep going, to burn off my uneasy energy, but Mary plunked herself down and pulled out our picnic.

"You don't seem sick," she said, setting out apples and the wraps she'd filled with leftover mole, noting that I had a pretty darn good appetite. "You just hiked two thousand feet in two hours, and you're chomping at the bit to keep going."

"But I'm not sure why I'm usually tired," I said, feeling compulsively for lumps in my armpit.

Mary insisted I take some of the oatmeal cookies she knew I loved. She told me to pull on my wool hat because I was shivering and chatted about how smart we'd been to pack extra fleece, steering me away from talking about the lump, her eyes low. I sat behind her and wrapped my arms and legs around her, and she leaned into me, and I whispered, "Thanks." By the time we returned to the trailhead, we'd put eighteen kilometers under our belts and sung ourselves dry. I would put distressing thoughts out of my mind while I waited for the biopsy results—only a few more days. Besides, the carbon-13 mass spectrometer data from the greenhouse experiment I'd conducted with Yuan Yuan earlier in the year would soon arrive, our test of whether the firs were communicating their stresses to the ponderosa pines, and I was eagerly awaiting the results.

Not to mention that I had two courses to teach. Plus five new graduate students and a postdoc to tend to, whose work would revolve around the principal question of my research program: How do mycorrhizal networks affect the regeneration of trees in our changing climate?

We drove straight to the pub, and Mary brought pints of stout to the deck overlooking the glacial blue Squamish River. The snowy peaks of the Tantalus Range stood silhouetted against the sinking sun. She clanged her glass against mine and said, "Sláinte" in her best Gaelic accent while k.d. lang's voice drifted like velvet from inside the bar, and I moved my chair closer to hers. She grabbed my hand and gave me her isn't-it-great-that-we're-up-to-no-good grin, then threw her head back to soak up the fading rays while I looked upstream at an osprey landing on a thicket-sized nest. But dread surged through me.

THE NEW DATA were on my screen.

I was floored.

The Douglas firs that Yuan Yuan and I had infested with western spruce budworm had dumped half of their photosynthetic carbon into their roots and mycorrhizas, and 10 percent of it had traveled straight to their ponderosa-pine neighbors. But what had me banging out an email to Yuan Yuan, now a professor at the Fujian Agriculture and Forestry University, was that only those pines connected by a mycorrhizal network to the dying firs, not those whose connections were restricted, were recipients of this inheritance.

Before I hit send, I looked out my window at the Pacific Ocean. A bald eagle drifted in to land on a shoreline Douglas fir, a silver fish wriggling in its beak. The week had passed, and I still hadn't received a call from the doctor. I checked my voicemail again and thought maybe no news was good news.

I reread the data, my eyes scanning the columns, and I whispered to myself, *"Saint chats!"* I sent my email to Yuan Yuan, sat back, and grinned.

This was a triumph a full year in the making—and now an answer was here. I'd jumped at her suggestion that we work together, already having mentioned her work in one of my review articles, and I was discussing her discoveries in my courses. She'd boldly advanced our knowledge on mycorrhizal networks, whizzing past historical hand-wringing over what constituted an interplant connection by inoculating her lab-grown plants with a plenitude of networking hyphae. Some scientists had still been puzzling over whether linking into networks affected the welfare of the recipient plants, and she'd gone well beyond this. She'd not only examined the receiver-tomato growth responses but had measured the activity of its defense genes, its production of the defense enzymes, and its resistance to disease. She was a gutsy free spirit who'd published her tomato experiment in *Nature's Scientific Reports*. I'd written back with an idea that had been germinating since the insect outbreaks had transformed our forests into oceans of dead trees: If dying trees communicate with incoming species, we might use this knowledge to better assist the migration of tree species as the old forests become maladapted to their native places. A warning-and-aid system—those infested Douglas firs telling pines to upgrade their defense arsenal, for instance—might be important for the growth of the new species or races (genotypes) as the old forests were dying back.

As the injured Mother Trees slowly folded their cards, did they transmit their remaining carbon and energy to their offspring? As part of the active dying process. Like senescing grasses, handing over their remaining photosynthates to boost the next generation. Maybe they just dispersed the content of their dying cells randomly to the rest of the ecosystem, since energy is neither created nor destroyed.

If all this could be revealed, we could better predict how tree species would migrate northward or upward in elevation as temperature warms—that is, to locations that would better fit their genes. As climate heats up, the forests will become sick and many trees will die, as was already happening, but new species preadapted to the warmer conditions should move in to take their place. Likewise, the seeds of the species in the dying forest should disperse into new areas that now matched their genes. One of the problems with the predictions was that this assumed trees would migrate at record speed—over a kilometer per year, rather than the fewer than a hundred meters we'd seen in the recent past. But the other assumption that seemed prevalent was that trees would move into purely empty spaces, as if the old forest had completely died. As if into a weeded clear-cut, where the new wave of planted trees would establish on a clean slate, unimpeded by any elders, as if they had packed up and left, even swept the floor, making way for the new. But this didn't make sense to me. At least some of the old trees—the legacies of the past forest—would remain. Not all of the trees would die, as Mary and I had seen at Tam McArthur Rim. These legacies should be crucial for helping the immigrants become established, maybe by including them in their mycorrhizal networks for an early boost of nutrients, or by providing shelter against the burning sun or summer frosts.

When Yuan Yuan had arrived a year earlier, in the fall of 2011, we'd collected buckets of topsoil from the Douglas-fir and ponderosa-pine forests near Kamloops and had already developed our experimental design over email so she could hit the ground running. We'd laughed a lot driving over the coastal mountains to the dry interior forests to collect the soil, her chuckle deep and gritty. Our kinship was immediate, maybe because of the common challenges of being female scientists and because of our shared interest in networks among plants. I admired her drive to get right to work, her passion to find answers, her eagerness to get her hands on a shovel.

At the university greenhouse, we placed ninety one-gallon pots on benches and filled them with the forest soil. In each pot, we planted one Douglas-fir and one ponderosa-pine seedling, but to alter the degree to which pine attached to the mycorrhizas of the fir, we planted one-third of the ponderosas in soil-filled mesh bags with pores large enough for any mycorrhizal hyphae, but not any roots, to grow through. We planted another third in mesh bags with pores so fine that only water could pass between the fir and pine. In the final third, we planted the ponderosas directly into the bare soil so their mycorrhizas could freely connect with the firs, roots intermingling. Our plan was to infest one-third of the firs in each of these soil treatments with western spruce budworms, snip the needles off another third with scissors, and leave the remaining third untouched as a control. Nine treatments all told, the soil and defoliation treatments fully crossed, with ten replicates of each.

We'd waited, Yuan Yuan increasingly nervous as her five months sped by, hoping the seedlings would be mycorrhizal enough to finish the experiment before her visa ran out.

After four months, we'd checked some seedling roots under the dissecting microscope, Yuan Yuan panicky when I said they looked bare. I then took some thin cross sections, squashed them on a slide, and looked under the compound scope: there were Hartig nets. The roots of Douglas firs and ponderosa pines had both become colonized with the single mycorrhizal fungus *Wilcoxina*. This told us that the fir and pine were connected by a *Wilcoxina* mycorrhizal network, except in the fine mesh, and we could go ahead with our defoliation treatments.

Yuan Yuan raced to the bug-rearing lab and grabbed her wriggling budworms. I ran to the mycorrhizal lab for scissors and sterilizing alcohol. We went into the greenhouse together to divest the seedlings of their needles. She put a breathable bag over a third of the seedlings and inserted in each a couple of budworms to feast on the needles. I clipped the needles off another third, leaving a few sprigs to carry out photosynthesis. The remaining third we left untouched.

One day after the defoliation, we put gas-tight plastic bags over the firs and pulsed them with $^{13}C\text{-}CO_2$. Another wait, with us imagining the sugar molecules traveling through the networks like milkshake through a straw. I called home that night, and Nava was excited to be practicing pointe for ballet, Hannah getting down her hip-hop steps. Their dance performance was a few months away, on Mother's Day, and I couldn't wait to be home. The next day, Yuan Yuan and I sampled the pine needles, doing the same the next day and the one after that, to check for defense-enzyme production. After six days, we pulled up all of the seedlings, ground them to bits, and sent the samples to a lab with a mass spectrometer to see if the firs had sent the pines any carbon isotope through the fungal links.

Now here we were, a few months later, looking at the data, Yuan Yuan in Jinshan while I was in Vancouver.

"See how much more carbon is flushed to the roots of the fir the more severely it's defoliated?" I emailed to Yuan Yuan, each of us on opposite sides of the world, united by the spreadsheet. "Yeah, I thought we'd find that," she typed back, explaining this was a well-known behavioral strategy for helping the attacked trees survive subsequent defoliations. A few minutes later, she added, "But I've never seen carbon after a defoliation migrate into the shoots of a neighbor." With defoliation, the firs became a large source of carbon, and the rapidly growing pines had drawn the carbon straight into their leaders.

"This matches with the defense-enzyme data," she tapped, sending a graph five minutes later. The firs had increased defense-enzyme production with the budworm infestation, which was normal, but within a day, the ponderosa pine *had done the same thing*. "But look,"

I wrote, "none of this happens unless the two tree species are linked in a network."

Yuan Yuan's email pinged in my in-box, "Wow!"

The pine's defense enzymes—four of them—had dramatically increased in perfect synchrony with the carbon dump, and this occurred only if the pines were linked belowground to the firs. Even slight injury to the firs elicited an enzyme response in the pine. The firs were communicating their stress to the pines *within twenty-four hours*.

What the trees were conveying made sense. Over millions of years, they'd evolved for survival, built relationships with their mutualists and competitors, and they were integrated with their partners in one system. The firs had sent warning signals that the forest was in danger, and the pines had been poised, eavesdropping for clues, wired to receive the messages, ensuring the community remained whole, still a healthy place to rear their offspring.

It hit me clear as spring water: I needed to be close to my children despite my fear, just as the dying trees were to theirs. I logged off and felt for the lump, smaller since the biopsy. I called Mary, back in Oregon, who'd been about to phone me herself.

"I need to go home and tell my girls," I said, explaining how Yuan Yuan and I had figured out that the dying firs were passing their carbon to the pines, and I'd pieced together that the dead Mother Tree I'd seen had done the same; this was how her two-year-olds had survived the drought. This was a cue to give my daughters my love, pass them all I could, in case I was dying too. I should rush to do this now, to make up for the time I'd missed being with them, when I'd been commuting to work.

"Slow down, you're not making sense," Mary said, as I prattled on that the data were telling me to get home and prepare them for what might be coming. "And you haven't heard from the doctor yet."

She offered to fly to Nelson, to be with me when I told them. They loved her blunt sense of humor, her modesty, her ability to fix things. Once she'd brought in her tools and tightened the screws on all of our wobbly chairs in an hour. They could count on her to tell them straight up what was going on. But I needed to go to them on my own, to let us absorb this as freely as we could together.

I told her she'd just arrived home herself, and I wanted her to rest.

Robyn offered to show up after her teaching so we could all celebrate that my lump was benign. She was optimistic. "Get home," she said. Be near the girls, keep things stable, show them I was calm.

I told the head of my department that I'd return the following week.

It had been almost two weeks since my biopsy. If I didn't hear anything tomorrow, I'd call the doctor's office. I asked Mum to come over to be with me when I phoned, and my first cousin once removed Barbara, the nurse, would drive over from Nakusp for good measure.

As I drove over the mountain passes, I felt compassionate toward the dying forest, harmonized, tied into the beauty of its hard wiring that passed wisdom to the next generation, as Grannie Winnie had done with me. But the infested trees were being clear-cut, the dead trees salvaged for the market. I wondered if, in our eagerness to turn a dollar, we'd cut off the opportunity for the dying trees to communicate with the new seedlings.

Hannah and Nava were waiting when I arrived with a pizza, and Don was there too. I grabbed my daughters in my arms and kissed Hannah's forehead, then Nava's. Hannah showed me her new biology kit and said she loved her teacher, and they'd already been talking about forest ecology. Nava did an arabesque, then took my hand and leaned in for a penché, and said they were choreographing a dance to "White Winter Hymnal" for the spring show, and they'd be wearing blue dresses and flowers in their hair. We leaned against the kitchen counter, eating pizza, with Don talking about the new snow on the mountain peaks, how winter was going to come fast this year, how great the skiing would be. The girls ran up to their new double beds to listen to their iPods, and I kicked myself for not having sat them down right away to tell them. "I know you're worried, Suze," Don said as they scampered up the stairs, "but you've always been healthy. I'm sure it's nothing." He stood with his hands in his pockets, his smile easy. He always knew how to reassure me.

"Thanks, Don," I said, glancing away.

My face scrunched up to stop tears as he put his boots on, and he gave me a hug. "Look, I know you. You're strong as an ox, and you'll get through this, no matter what. But call me when you hear." We stood a moment, feeling strange about the new order of things, before he grabbed his coat and disappeared out the back door, his taillights fading down the alley. I climbed the stairs with the rest of the pizza. The girls and I stayed on Nava's balcony watching the sunset over Elephant Mountain, the tinge of snow glowing in the pink light.

When it got too cold to sit outside, we sat on Nava's bed and I told them that I'd had some tests and would be getting the results tomorrow. They looked at me wide-eyed, but I added, "No matter the outcome, I want you to know that I'll be fine, we will all be fine."

Hannah asked how they take the test, and Nava asked what breast cancer was, and I told them what I knew, and how they would have to have checkups when they got older too. All women need to look after themselves like that. They hugged me, and I told them I loved them, and by the time I kissed them good night, I was feeling a little lighter.

Friday morning, the girls walked to school, and I raced up my favorite mountain trail, wanting to put off the phone call for another few hours, speeding through the open forest, the ponderosas giving way to the Douglas firs and aspens and lodgepole pines farther up. The October hoarfrost on the path was frozen in icy feathers, and on my way to the knoll, I slipped past two brown bears gorging on ripe huckleberries. At the top, I called Mary and told her I was ready to speak to the doctor. I made a wide detour around the bears on my way back down, all of us acutely aware of the hierarchy of strength. The vanilla aroma of the ponderosas filled me. *Every little thing's gonna be all right.* I imagined the trickle of water percolating through the fungi linking the ponderosa pines with the firs, the lodgepoles with the aspens. I was at home with these trees, my gentle friends, my confidants.

Mum was coming down the alley, carrying cups of coffee, her white hair glinting, red gum boots and worn coveralls on for the garden. Barbara appeared with a pot of hamburger stew, a tea cloth on top. They sat on the bench on my porch, sipping coffee, and my home phone began ringing. I walked inside to pick it up and brought the receiver outside. Mum and Barbara abruptly stopped chatting and looked at me through the steam off their cups.

I listened to the doctor. Tests and options and so many words that didn't register. I thought of the Mother Trees providing shelter and nutrients and shade from the sun, protecting and caring for others even as they diminished. I thought of my daughters. My beautiful, precious daughters who were brilliant flowers growing and blooming.

I closed my eyes.

Not even Mother Trees can live forever.

14

BIRTHDAYS

"Here are some survivors," said Amanda, my master's student, while crouching at the dripline of the Mother Tree. It was late October, halfway between Kamloops and the rodeo grounds where I'd watched Kelly thirty years ago, and snow was falling as gently as a baby's breath.

The Douglas-fir Mother Tree looked as though she'd been through a knothole sideways, crown ragged from the felling of her neighbors and her trunk scarred from a skidder backing into her, but she had produced plenty of cones this past summer. Black-capped chickadees loved her for this and hopped along her branches. I admired her determination to carry on, to care for her young in spite of the shock of her losses. My mastectomy was to be in a month, and the treatment to follow depended on whether the cancer had spread to my lymph nodes. Barbara had advised me to try not to spin scary what-if scenarios. The publication of papers I'd written with my lab describing the architecture of the mycorrhizal network, and my conceptualization of Mother Trees, and the heartwarming public responses to our *Mother Trees Connect the Forest* film were helping. One highly respected scientist wrote to me that the discovery will "forever change how people view forests." It helped being out here with the Mother Trees too.

I'd been mulling over the possibility that kin recognition, something we normally ascribe to humans and animals, might occur in Douglas fir. I'd jot down notes when stopping for gas after stretches on the road, along with lists of the things I still hadn't done. The idea didn't just arise from late-night fatigue at the wheel; someone else had put it in my head. I'd read a paper by Dr. Susan Dudley of McMaster University in Canada about her discovery that an annual plant—the searocket, *Cakile edentula,* of the Great Lakes' sand dunes—could distinguish between neighbors that were kin (siblings from the same mother) and those that were strangers, from different mothers, and that the cues came through their roots. Winding my car around the cliffs in the moonlight, I wondered if conifers could detect relatives too. A Douglas-fir forest was genetically diverse, with wind-pollinated kin and stranger seedlings establishing around Mother Trees. *Could Mother Trees distinguish kin from stranger seedlings?*

Since we'd discovered that Douglas-fir seedlings grew up linked into the mycorrhizal network of old trees, I'd figured if kin recognition were occurring, and if it involved root cues as Susan had found with her searockets, it would have to be signaled through fungal links, because all tree roots were coated in mycorrhizal fungi. Also, given that populations of Douglas fir were regionally distinct, with less genetic variation in local valleys than across mountain ranges, there should be lots of relatives in close proximity to Mother Trees. If relatives had lived closely together for centuries, surely, I thought, there must be a fitness advantage to recognizing one another. To helping one another along, carrying forward the family line. Maybe the Mother Tree could alter her behavior—make some elbow room—to increase the fitness of her kin. Or transmit nutrients or signals to her offspring. Or even shoo them away if the soil was not favorable. That is not to downplay the crucial role of maintaining genetic diversity to ensure the forest is adaptive and strong and resilient. In that varied gene pool, though, there could also be some role for old trees in shedding their locally adapted seed and nurturing their kin.

I'd always been willing to push the envelope, but I'd become more relaxed as a scientist in recent years, my articles on mycorrhizal networks getting more favorable reviews lately. I didn't know why. Maybe it was that more studies were corroborating my original findings that birch and fir shared carbon, or maybe I was just at that stage of my career where I was better known. Whatever it was, I'd been enjoying the freedom to ask riskier questions. And Amanda had been happy to go along with me. "It could be a wild goose chase," I'd said, warning her that the possibility of Douglas-fir Mother Trees recognizing their kin was low and we could turn up nothing, but at least she'd learn how to do an experiment.

"Well?" I asked as we examined three petite green parasols ringed within the fringe of the lunch-sized mesh bag she'd dug into the soil six months earlier. Amanda, five foot nine, strong from playing on the national baseball and hockey teams, undeterred by the snow, checked another bag. She pointed to a cluster of red seedlings and said, "Many of the kin are alive, but the strangers are dead." Unrelated and unconnected to the Mother Tree, the strangers had perished in the summer drought.

We walked toward the other fourteen Mother Trees, left by the loggers as habitat for wildlife, as my thoughts slid into a dark corner. A friend told me that a powerful colleague had said to him, "You don't believe that trees cooperate with each other, do you?" Something I might expect from old-school foresters but not from the halls of academic freedom. The thirty-year battle over the entrenched dogma that competition was the only interplant interaction that mattered in forests was getting the better of me today.

I followed Amanda over logs and through puddles to the next Mother Tree, her branches dusted with new snow. She asked if I wanted to rest; it would be understandable. I stammered, "I'm all right" but dropped onto a stump to take notes while she kept checking the bags. Under this Mother Tree, just as with the first, more kin seedlings were alive than strangers, especially in bags that allowed them to connect to the network. I chewed the end of my pencil. It was possible birch also sent more carbon to related birches than firs in mixed stands, but I hadn't tested this in my doctoral research. And dying firs may send more carbon to other firs than to pines, as my experiment with Yuan Yuan had shown, but the fir-fir pairings we'd planted hadn't established well enough in the greenhouse to run that test. One of my graduate students had shown that Douglas-fir Mother Trees facilitated establishment of fir seedlings, but at that time we hadn't thought to test whether she favored related seedlings over strangers. It made sense evolutionarily that a Mother Tree would favor her own, regardless of her species identity.

Amanda had started her master's a year earlier, in the fall of 2011,

asking—after we'd published the map of the mycorrhizal network—
the next obvious searocket-like questions about Mother Trees recognizing their young and granting special favors. I already knew Mother
Trees shared resources with strangers, since my students and I had
examined this extensively before I'd learned of Dr. Susan Dudley's
work. If there were kin recognition from Mother Trees, especially
via mycorrhizal networks, was it expressed in fitness traits—did kin
grow bigger or survive better than strangers? Or perhaps recognizing
kin was expressed in adaptive traits, such as root or shoot growth.
Amanda was testing these questions in this field experiment and in
two greenhouse trials at the university.

I rested while Amanda checked more mesh bags. In the spring,
she'd installed twenty-four of them around each of the fifteen Mother
Trees in this clear-cut. Twelve of the bags had pores big enough for
the mycorrhizal hyphae of the Mother Tree to grow through and colonize the germinants. The other twelve had pores too small to form
a network. In each of these mesh-bag types, Amanda had sown six
with seeds from the Mother Tree (kin) and six with seeds from different Mother Trees (strangers). The four treatments—the two mesh
and two relation treatments fully crossed—were applied at each of the
fifteen Mother Trees, a number meant to give us confidence in any
trends. To ensure our findings were not an anomaly of this location,
we'd repeated the experiment at two more sites. This clear-cut near
Kamloops was the hottest and driest, and the other two farther north
were cooler and wetter.

For sowing kin seeds, Amanda had collected cones from our total
of forty-five Mother Trees the previous fall. She'd used pruning shears
where the Mother Trees were fewer than ten meters in height but
hired a girl to use a shotgun where they were taller. I imagined the
Winchester on her shoulder, barrel aimed high, the deafening pop,
branches and cones tumbling, the squirrels scampering for cover, eyes
on the jackpot. Over the winter, we hired slews of undergraduates
to open the cone scales, collect the seeds, and test whether they were
viable. In this particular year, the climate hadn't aligned for fir all that
well, and many seeds had been dead.

We reached the last of the Mother Trees on this site, and Amanda brushed snow off a stump for me. She poured some tea, and its steam warmed my hands and face while she methodically checked each of these final bags, calling out the number of survivors.

My phone rang. Mary had made it home, and she'd be back as soon as she put her plants to bed for the winter. She'd rushed to Nelson after my diagnosis. When I told my extended family that same day I had a girlfriend, Mum simply said that she was glad I had someone. I was proud of my family for accepting this, for being fine with who we all were.

The snow was coming down harder. Even before adding up the numbers, I knew Amanda and I had gone beyond confirming that Douglas-fir seedlings tended to perform better if linked to a healthy, unrelated Douglas-fir Mother Tree: seedlings that were her kin survived better and were noticeably bigger than those that were strangers linked into the network, a strong hint that Douglas-fir Mother Trees could recognize their own. I suggested following these seedlings another year.

"I'd feel better if we could," Amanda said, stowing the notes in her pack. She liked this experiment, her first one ever, and I guessed she'd keep coming back as long as her seedlings were still alive. In the agreeable shelter of this Mother Tree, the struggle was worth it.

JEAN JOINED ME in Vancouver for an InspireHealth workshop on surviving cancer. Experts walked us through ways to improve our chances—by exercising, eating well, sleeping well, and reducing stress. But the most important thing was to ensure that our relationships were strong and to keep communicating how we felt. *We are defined by our relationships,* a doctor said. The single characteristic of people who survive cancer: they never give up hope.

Mon Dieu! C'est ça! I thought, *this is something I can work on.* I was still so introverted, sensitive, stumbling too easily over what others thought. I'd been too agreeable when a forester had said to me, "I want to cut these Mother Fucking Trees down because they're going to blow over anyway, and we might as well make money." I was still afraid to stand strong with my convictions, fight tooth and nail. But isn't this what my trees were showing me too? That health depends on the ability to connect and communicate. This cancer diagnosis was telling me I needed to slow down, grow a backbone, and speak out about what I'd learned from the trees.

The surgeon removed both my breasts, and I woke up to Mary and Jean, Barbara and Robyn hovering as I looked at my flattened chest and pressed the morphine pump. A few days later, I was in my apartment, eating kale and salmon, my scars red and my bruises as purple as eggplant. I walked a hundred meters, then another and another, ready to go home to Hannah and Nava for Christmas. We just needed the full biopsy results. "You might be done with treatments if your lymph nodes are clean," Barbara told me.

On our way out of town, we learned that the cancer had spread to my lymph nodes.

The two oncology doctors, Dr. Malpass in Nelson and Dr. Sun in Vancouver, said I would receive a new "dose dense" regime of eight chemotherapy infusions, one every two weeks over four months, the most effective option for my kind of cancer. They figured I was young and fit enough to handle it. The first half would be a combination of two older drugs, cyclophosphamide and doxorubicin—what Barbara called "the Red Devil"—and the second was paclitaxel, derived from the Pacific yew tree. Dr. Malpass, lanky and compassionate, would care for me during chemo, and Dr. Sun, diminutive and quick to laugh, would take over in the time to follow. *I should have moved to Nelson to live a quiet family life,* I thought as they explained the possible side effects. There were the common ones—nausea, fatigue, infections. And the less common—stroke, heart attack, leukemia. Don was right—I should never have pursued the university job. And God knows I shouldn't have sprayed Roundup in those early experiments, or missed checking the safety latch on the neutron probe, or forgotten to press down the nose clips on the dust mask when I was grinding the radioactive seedlings. And all that stress over the breakdown of my marriage surely did not help.

A FEW WEEKS LATER, at the beginning of January in 2013, a nurse put a needle in my skin, and the cherry-colored Red Devil coursed through my veins. I imagined the cancer cells shriveling as I gazed out the hospital window at the snow falling on a lone tree. She stood sentinel over the hospital, the town below, and the ash, chestnuts, and elms lining the streets, trees helping trees, people helping people. Bring it on—if this tree could live with her roots cut off from the wild forest, I could beat this too. The next day, I skied twenty kilometers up my favorite trail, leaving Robyn and Bill in my wake, as though to prove I was tougher than cancer. I passed a clear-cut, the planted pines a meter taller than last year, and I thanked the trees along the boundary for helping the saplings along. "I need your help. I need to be healed," I said at the top of the trail—where they stood solid, still. I glided along, their branches over me, some touching my arm. The next day, I barely made it around the one-kilometer loop, my body a bag of wet cement, landing me plastered to the couch, Bill checking on me. He was a filmmaker, brilliantly creative but in a stretch of downtime, so he was here to help. He sat with me, patiently. Not much talk, no fuss, just there. After a week, the drugs settling into my cells, I was back on my skis, increasing to two, five, then ten kilometers, Bill trailing to make sure I was okay.

"LOOK AT MY PIROUETTE," Nava said, tipping onto her toes. I held her hand above her head, and she twirled like a spinning top. Hannah put on the glinting black-and-gold high-tops Grannie Junebug had given her and did some break dancing, tuts, and swipes. I tried a step, but my feet were numb. During their shows, dances beautifully choreographed, bodies precisely trained, my brimming eyes were always fixed on them, only them.

I was counting on chemo being finished by Mother's Day, the weekend of their grand finale concert, the annual spring showcase, but during my second infusion, Dr. Malpass showed me my chest X-ray. Cheryl, a veteran chemo nurse in a flowered uniform, looked at the screen in concern, while the other nurse, Annette, was patting the arms of patients hooked up to drip lines, asking how they felt. "I've never seen anything like this. Your heart has grown by 25 percent over the past two weeks," Dr. Malpass said, pointing at the X-ray, the portacath the surgeon had inserted below my right collarbone in sharp relief. My lungs, ribs, and heart were in clear outline of "before" and "after" images. *That's me—at least the new me,* I thought, running my hand over my chest, ribs like rulers.

"I see," I whispered.

"You might have had a heart attack," he said. "You'll need more tests—and please stop skiing so you can focus on fighting the cancer."

Hannah suggested I go walking instead. She was leaning into me that night while we watched *Glee.* My laptop was perched on a stack of books on the oak coffee table, the girls' homework abandoned. We were eating bowls of chickpeas, yams, and rice by the bay window. Elephant Mountain glowed across the lake. We watched the double wedding of Kurt and Blaine, and Brittany and Santana, Santana's grandmother finally accepting that two females can marry each other. I felt slightly embarrassed, but Hannah loved the scene, Nava too, and I thanked my lucky stars that kids are more accepting these days. "I can only walk on the flats," I said when the episode ended. I had never missed a season of skiing, and neither had they, having started as soon as they could walk, but Nava piped up, "The snow will be better next year anyway, Mama."

Two weeks after chemotherapy started, right before my hair fell out, January 2013

We'd persevere. We had to.

Mary came to help me through the next round of chemo, my heart given the all clear. A petite, kerchiefed seventy-year-old woman was in the chair by the window when we arrived. "She stole our spot," Mary whispered. We found a different one. There were four, one spot in each corner of the room, beige curtains providing the barest shades of privacy, the nurse's station in the center, a bank of picture windows along one side. The woman fiddled with her bag of pills, the same ones I'd become an expert at taking. Pink pills to reduce nausea, blue ones to manage thrush, the awful-tasting ones to keep my bowels moving. I slipped past the curtain to introduce myself. Her name was Anne, and her husband was in another room dying from heart failure.

The next day in the shower, I looked at my feet, and there was my hair. Like a wig in the rain. I touched my head, and the remaining hairs were escaping like seeds from a dandelion. When I passed the mirror, I couldn't look. "Let's go to the forest," Mary said, and I put on two warm hats, the first in place of my hair, the second to keep the wind from freezing my scalp, and we walked in the falling snow among the cedars, the young ones layered in circles around the old ones. "Of course," I whispered as we slipped past the saplings, thinking that seedlings could be intermediate nodes between distant Mother Trees and would eventually become Mothers themselves. This unbroken line between the old and the young, the link between generations, as with all living things, is the legacy of the forest, the root of our survival.

Mary would bring breakfast to bed each morning, read me a chapter of *The Unexpected Mrs. Pollifax,* and then take my arm so we could limp along the windy shoreline of Kootenay Lake. She cooked the salmon and kale, complaining that kale in Canada was tough as nails, then she'd sneak in a chicken pot pie and bowl of ice cream.

At my third round, Dr. Malpass asked me to speak with another patient. Lonnie and her sister, in their midforties, dropped by my chair to talk about her "dose dense" treatment, the same as mine. Lonnie clutched her purse, with one of those old-fashioned kissing clasps, as she stared at the tubes running into my veins. "It's not that bad," I said, though I was more tired with each round.

"I don't want to lose my hair," Lonnie said, looking at my tuque, her voice strained. What a rip-off that we lose that part of our identity when we need it most. I invited her to lie on one of my couches at home after her first treatment, and she accepted. She came back the time after that. Soon we were joking about throwing away our couches and clothes and hats and wigs once the chemo was done. Lonnie lived in the forest, half an hour out of town, and we sat on her chesterfield sometimes, looking at the trees and the snow hugging her house, wishing for spring.

"You should meet Anne," I said, and soon the three of us were texting back and forth.

I kept daily records, scoring from one to ten my fatigue, mood, fogginess—what they called chemo brain, the inability to put together thoughts, remember words, speak in sentences. My spirit tumbled with my energy: I was depressed in the days following chemo. Just walking around the block felt like swimming against a riptide, and I understood what the end of life felt like, that there simply wasn't enough energy to go another step. Death wasn't a bad option if you couldn't eat, go to the bathroom, or get off the couch. When you couldn't put on your skis to shuffle down the track along the river or make dinner for your kids. "I'm trying hard to be me," I wrote in my diary, wanting to be normal again, to ski with my girls. I'd pick up for a day, then tank again, then up, then down before I crawled back up—right before being dosed again. "You're riding the double dip," Dr. Sun said when I showed her my serpent graph.

At the fourth and last infusion of the Red Devil, I told Dr. Malpass I wasn't sure I could do any more. Even tears hurt. He suggested meditation, sleeping pills, and sunshine and promised I'd feel better in the last four rounds when we switched to the medicine of the yew trees.

Anne sent a text: "Think of what you want to be, not what you don't want to be." Strong like my trees, I thought, like my maple. That afternoon, I sat at her base, the swing immobile. I leaned my back against her, face to the warmth, and felt myself seep into her roots. Instantly I was inside my maple tree, her fibers interweaving mine, absorbed into her heartwood.

AMANDA'S KIN-RECOGNITION EXPERIMENT in the three clear-cuts was just the beginning. Since I couldn't let her master's degree depend on a field study that might be bound for failure, we'd matched it with a greenhouse experiment where she'd grown one hundred seedlings—what we called the "Mother Trees" for the purpose of this experiment—for eight months, then planted fifty of the pots with a side-by-side sibling and the other fifty with a stranger. With each type of neighbor—kin or stranger—twenty-five were grown in mesh bags with pores wide enough for signaling through mycorrhizal linkages, and the other twenty-five were grown in fine-pored bags where mycorrhizal networks could not form. We grew the pairs until the Mother Trees were a year old and the new neighbors were four months.

Back in March—right before my fourth Red Devil infusion—Amanda had emailed that she was ready to harvest her one hundred pots. "Before you do, you and Brian should label the Mother Trees with ^{13}C-CO$_2$, to see if they share more carbon with kin than strangers," I'd replied. Trapped inside my body, I'd been obsessing about the degree to which Mother Trees might not only identify with their own but shift the amount of carbon transfer in their favor. Brian was my new postdoc, having stepped in to help my grad students with their lab work and data analyses. "Don't worry, Suzanne, I've got this," he'd assured me in his British brogue. We'd have to time our Skype meetings according to my stamina. The day they labeled, I felt as if I were climbing a mountain without air, wishing I were in the game but grateful they were carrying on without me there. "We stayed up all night in the greenhouse," Brian wrote after the trees were harvested, their mycorrhizas counted, tissues ground for carbon-13 analysis. I lay back on the couch with a sigh.

A month later, we conferred on Skype, with Amanda's data tables and figures on the screen. She started by saying, "Hey, you look great."

"Yeah, thanks, I'm hanging in here," I said, tilting the laptop in the hope of hiding the bags under my eyes as she stickhandled me through the data. I'd been to one of her hockey games with Hannah and Nava, while her parents, Loris and George, and her auntie Diane, were cheering in the row behind us. Amanda was captain of the UBC women's team, fast on her skates, deft with her stick. She knew how to line things up, eye a goal.

"Kin neighbors have more iron than the strangers," she said, tracing her cursor over the differences between the two types of neighbors and then showing me the same with copper and aluminum. "The Mother Trees could be delivering these nutrients to their youngsters," I said, struck with a vision of her passing a puck sharply to the center player, who tore with it toward the goal and swiftly passed it to her wingman as Amanda took up defense at the blue line. These three micronutrients were essential in photosynthesis and seedling growth, I said, and we bantered about whether iron, copper, or aluminum could also be part of signal molecules that move from the Mother Tree to her kin.

Like passing a puck.

"Kin seedlings also have heavier root tips than strangers and greater colonization by the mycorrhizas of the Mother Trees," she said, her cursor hovering over the data points.

"Oh, that fits!" I said.

"Do you think it's important that we're seeing that Mother Trees are also bigger when they're next to their kin?" Amanda asked. "It makes sense if they're passing signals back and forth."

Of course it did. Being connected and communicative affects the parents as much as the kids.

The next day, I clicked on Skype to join Amanda and Brian looking over the isotope data. Even before their images grew into focus, Brian was excitedly saying, "Look at this!"

"The amounts are small," Amanda said, "but the Mother Trees are sending more carbon to the mycorrhizal fungi of their kin than the others! Kin-recognition molecules seem to have carbon *and* micronutrients." The arrow of a mouse skirted across the screen.

"Pure dead brilliant," Brian said softly, even if the carbon hadn't quite made it all the way into the kin-seedling shoots. I'd already observed carbon move from birch into the shoots of fir, and from dying firs into the shoots of connected pines, so I was surprised to see carbon transmitted from a Mother Tree stop short at the mycorrhizal fungi of her kin, not moving into those shoots. But Amanda's kin seedlings were only one-fifth the weight of the recipient pines in the fir-defoliation experiment with Yuan Yuan, and I guessed that, unlike the pines, Amanda's kin firs were still too small to generate a strong enough sink to draw the carbon into their shoots. Not only that, the source strength of Amanda's donor firs would be smaller than that of Yuan Yuan's dying firs, because they would be using most of it for their own growth and maintenance instead of dumping it into the network. *If I ever get through these damned chemo treatments,* I thought, *I'll have to test this again in a later experiment with dying Mother Trees and larger kin.*

"Even a tiny amount moving into the mycorrhizal fungi of the seed-lings could mean the difference between life and death when the little ones are small," I said. Germinants struggling to survive in the deep shade, or during the summer dry spell, could live instead of die with the slightest boost, the smallest of advantages, if it came at the right time. Not only that, the bigger the Mother Tree, the healthier she was, the more carbon she gave.

There is even more here, I thought as I logged off. In my kitchen, frost was creeping across a pane. I was looking forward to Mary's visit, and Jean's. Communication among relatives is important, but it also matters in whole communities. In a couple of experimental families, the Mother Trees even gave as much to the mycorrhizas of a stranger as her kin. Of course, not all families are the same. Forests are mosaics too. That's what makes them thrive. Birch and fir transmitted carbon to each other, even though they were different species, and to the cedars in their unique arbuscular mycorrhizal network. These old trees were not only favoring their kin, they were also ensuring the community in which they were raising their kin was healthy.

Bien sûr! Mother Trees give their kids a head start, but they also tend the village to ensure it flourishes for their young.

Amanda and I scoured her field data. Only 9 percent of the seeds had germinated at the three clear-cuts. I recalled sitting on the logs taking notes as she'd checked the bags, no idea yet what being tired truly meant. But from disasters sometimes glitter gold, and I'm not one for tossing aside a trend that looks intriguing.

"The correlation is weak between the number of kin establishing and the aridity of the climate," Amanda said, almost apologetically, "but I did see the same trend in the greenhouse experiment." Kin appeared more dependent on Mother Trees in the dry rather than the wet climatic areas. The Mother Tree had especially stepped in to help at the driest site, perhaps by transporting water to her seedlings through the network.

Glasses of half-finished club soda cluttered my table as I wrote in my diary. My energy was at five, mood exceptional today. Maybe society should keep old Mother Trees around—instead of cutting most of them down—so they can naturally shed their seed and nurture their own seedlings. Maybe clear-cutting the old, even if they're not well, wasn't such a good idea. The dying still have much to give. We already knew the elders were habitat for old-growth-dependent birds and mammals and fungi. That old trees stored far more carbon than young ones. They protected the prodigious amounts hidden in the soil, and they were the sources of fresh water and clean air. Those old souls have been through great changes, and this affected their genes. Through the changes, they'd gathered crucial wisdom, and they offered this up to their offspring—providing protection, laps into which the new generations got started, the foundation from which to grow.

The door slammed. Nava and Hannah were home from school, tuques covered in snow. Hannah needed help with her math, and we opened her books.

My unfinished business—my main lingering question—centered on this: Were the old Douglas-fir Mother Trees that were unhealthy—sick with disease, stressed from the drought of climate change, or just ready to pass on—using their last moments to transfer their remaining energy and substance to their offspring? With so many forests dying, we should figure out if the elders leave a legacy. Yuan Yuan and I had already seen that stressed firs passed more carbon to neighboring pines than did healthy firs, and Amanda had also discovered that, in the proximity of healthy Mother-Tree seedlings, kin seedlings had better nutrition than strangers, and their mycorrhizal fungi received more carbon. *But so far we hadn't seen whether dying Mother Trees passed their carbon legacy into the shoots, the lifeblood, of her kin seedlings, beyond the fungal web.* Therefore we couldn't verify that the carbon transferring into the fungi actually improved the kin seedlings' fitness. We didn't know if the fungus kept the carbon for itself, like a middleman, or if the carbon sent by the Mother Tree was truly used to increase the survival of her children.

And if the urgency of death compelled a Mother Tree to funnel even more of her substance into the photosynthetic machinery of her children, that would have implications for the entire ecosystem.

It would take years for the full answer to arrive. But first I had to inch up the hospital stairs for the start of my paclitaxel infusions.

A drug derived from the yew tree.

"YOU HAVE GOT to pull yourself together for Nava," Robyn told me, trying to hide her worry. I was staring at the presents I needed to wrap, my portacath riddled with needle marks, my throat white with infections, my bald scalp itching. The salami sandwiches I was trying to prepare for the birthday party were making me gag. My drugs were piled on the curio cabinet, along with Mary's chart for tracking my slew of pills. The needles for injecting filgrastim into my stomach were on display to remind me of my nightly ritual. My mouth tasted like shit—literally. The nausea wasn't as bad with the paclitaxel infusion, but the fatigue was worse. I was having a hard time enjoying what meant the most to me: time with my daughters.

"I can't."

"Yes, you can," she said quietly. She finished the sandwiches and wrapped them in waxed paper.

Robyn had moved in these past weeks while Mary was away, sleeping in the hallway outside my bedroom, waking with my every moan. She came right after teaching first grade each day to make dinner.

Nava peeked around the door. Today she was thirteen. She had put on her favorite dress, maroon with pink flowers, a reminder that March 22 was the first day after the arrival of spring. Five friends were due in an hour at Lakeside Park, a few blocks from our house. She listed her sea-green eyes toward me and asked if it would still be all right to have her party.

"Oh, sweetie pie." I pulled myself up from my chair. "I'll be at the park in a jiffy."

Sandwiches, fizzy drinks, chocolate cake. I pulled the wagon with the party food and balloons to the picnic table. Snow lay in patches, and the limbs of the maples and chestnuts were bare, the roses covered in burlap, but the sand down to the water was full of footprints. Han-

Nava on her thirteenth birthday, March 22, 2013

nah and Grannie Junebug arrived while Auntie Robyn was setting
out yellow napkins and cups—Nava's favorite color—and insisted the
birthday girl open her present, an aqua mug with "Nava" in black
letters. Grannie Junebug placed a tiny box in front of Nava and said,
"Grannie Winnie gave me this watch when I was thirteen. I want
you to have it now." Mum gets it just right sometimes. Nava tried it
on—the oval face was inlaid with pearl, and the band was made of
interlinked gold and silver hearts.

The paper plates were printed with ballerinas. The girls ate their
sandwiches and drank orange soda, dyeing their lips, and we put the
candles on the cake—with "Nava" in yellow across the chocolate frost-
ing. I used to organize treasure hunts for their birthdays, with elaborate clues and mazes and prizes. Today Hannah proposed an egg race,
and she had brought a carton and six spoons. At her urging, I lined
up the girls, each with an egg perched on a spoon, and shouted, "Go!"
Off they sprinted toward the finish line, all of them laughing and
dropping their eggs with a splat, including Nava.

A breeze blew up from the lake, the first sailboat of the year jibing against the frigid air, bare limbs of the aspen clones rising white,
crowns of the birches glowing red, the branches of the ponderosas and
fir stretching dark in anticipation of spring.

I stuck candles in the cake, fumbled with the match, and cupped
my body to keep the wind from blowing out the flames. "Make a

wish!" Auntie Robyn said, and as Nava inhaled, I made a wish too, for the good health of all of us, and that I'd be with my trees again soon, and we all blew, just to make sure. A last flame flickered until a breeze snuffed it, and we sang "Happy Birthday." A whiskey jack hovered. Her smile as wide as the moon, Nava said, "Thanks, Mum." I whispered, "Sweetie pie, you have the whole world in front of you." Just as I too felt born into a new lease on life, a fresh spirit rescuing me. I turned her shoulders with my hands, and she launched into an elegant twirl, spinning five times in a chaîné, her eyes meeting mine with each turn. She tapped my fingers one last time before she let go.

I resolved that I *would* be here for the graduations of my children. On April 22, Hannah would turn fifteen. Earth Day. Nava born at the start of spring, the day we're meant to pause and think of land, sea, birds, creatures, one another—how could I fail to enjoy the whimsy of that, the uncanny timing of my bringing them into this world?

By autumn of that year, I would venture from my kin circle into a nurturing of other children. In New Orleans, despite ongoing exhaustion, I'd give a TEDYouth talk to one hundred fourteen-year-olds perched on beanbag chairs. I would practice with Mary so I could do a good-enough job for the film to get posted on YouTube, Mary patient with my rehearsals ad nauseam, providing little mnemonics until I could link my sentences despite a course of radiation after chemo that still made my brain shutter. I struggled with anthropomorphisms that I knew would be criticized by scientists, but I chose to use terms such as "mother" and "her" and "children" anyway to help the kids understand the concepts. The host would turn out to be colorful, his animation the antidote to my introversion. Seven minutes of talking about the importance of connection while standing in front of Bill's beautiful images of trees and networks, and the host was on his feet, overjoyed. The video did go online, and received more than seventy thousand hits, and I was invited to appear on the main TED stage two years down the road. I was happy that my recent work was being well received, and citations of my handful of review papers had exceeded a thousand.

LONNIE, ANNE, AND I crowded around Denise at the cancer support group not long after Nava's party. On her first visit to the chemo room, Denise had fled in tears because I looked almost dead in the chair, and she thought that would be her soon. Anne, Lonnie, and I were already in a web, trading texts about our aches and fears, giving one another lucky rocks and poems, sharing information about this cream or that potion that might get rid of a sore throat or rash. Anne would text us: "Your body will follow your thoughts, so think healing thoughts." She had become our Mother Tree as we trudged through our final infusions.

Denise joined us for lunch, instantly part of our sisterhood. At my round table, Lonnie set out her borscht, Denise the gluten-free crackers, me the kale salad. Anne contributed dark chocolates, saying she couldn't follow *all* the rules. My thrush was acting up, and Lonnie wasn't sleeping well, and Denise's feet were numb, and Anne reminded us we were almost finished with chemo. "Eyes on the prize," she said. The true prize, we all knew, was that we were together, a friendship melded out of devastating diagnoses and hardship, facing death as one, never letting one another give up, picking one another up when we couldn't take another second. I knew then, my connections always strong, that even in death I would be okay. Lonnie asked if her blond wig looked better than her original hair, and we shouted yes.

"Let's give ourselves a name," said Lonnie. "The BFFs, Breastless Friends Forever." "But I still have boobs," Denise said.

I said her lumpectomy still qualified her.

A week later, after her third infusion, Anne met me on her way out of the chemo room while I was going in. "My poor Dan is going to pass soon," she said, fidgeting with her scarf, then patting my arm before I could get the words out that I was sorry.

A few hours later, she texted that Dan had died in her arms.

DR. MALPASS WAS RIGHT. The paclitaxel infusions were easier to absorb than the earlier chemo drugs, and I regained some energy and started to walk in the forest again. Paclitaxel is derived from the cambium of the yew—a short, shrubby tree that grows under old cedars and maples and firs. The Aboriginal people knew its potency, making infusions and poultices to treat illness, rubbing its needles on their skin for strength, bathing in preparations to cleanse their bodies. They used this tree to make bowls and combs and snowshoes, and to craft hooks and spears and arrows. When the anticancer qualities of the yew were brought to the attention of the modern pharmaceutical industry, there was a bounty on the trees. I'd find the small yews—their branches as long as their stems—stripped naked of their bark, looking like crosses, specters of maltreatment. In recent years, pharmaceutical labs have learned to synthesize paclitaxel artificially, leaving the yews to thrive under the cool canopy of the forests. When the old growth is clear-cut for the big timbers, however, these small scaly trees are left weakened in the hot sun.

When Mary arrived, we went seeking the yews, finding them in the shifting shadows of the cedars and maples, their branches luxuriant, bark medieval and shaggy, their stature no greater than that of a hobbit. Where their lower branches had touched the ground, new stems had taken root and entwined around the Mother Trees. I ran one of her branches through my hands, the rows of needles in pairs, dark green above and gray green below. Her skin felt like silk, even though she was elderly, her most aged relatives in England thousands of years old. I tugged at her bark to say hello, and it peeled off in my hands. The cambium below glowed purple.

After the last of the paclitaxel had been shot into my veins, I brought Hannah and Nava to this grove. The spring beauties and skunk cabbages were in bloom. "These are the yews that made my medicine," I said, and we put our arms around their gnarly trunks. I asked them to look after my daughters, all of the daughters, as they did me. In return, I promised to protect them, ask questions of them, seek treasures still unknown. Unlike most conifers here, they formed relationships with arbuscular mycorrhizas, so did they connect with the cedars and maples? I bet that they kibitzed with the bigger trees and the tiny plants at their roots—the wild ginger and rosy twisted-stalk and false lilies of the valley. A thriving, webbed neighborhood might boost the yew in producing paclitaxel in greater abundance, stronger potency.

What am I if I don't give back?

I pictured roaming among the yews when I was better, smelling their sharp sap, working with them in the shade. I told this idea to my daughters, and we walked among the cedars and maples towering over the yews, and Hannah said, "You should do that, Mama." We ducked under the crowns of the Mother Trees and ran through their circles of young. Nava untwirled the scarf Mary had given her and wrapped it around the oldest one, her branches so long they touched the ground.

Our modern societies have made the assumption that trees don't have the same capacities as humans. They don't have nurturing instincts. They don't cure one another, don't administer care. But now we know Mother Trees can truly nurture their offspring. Douglas firs, it turns out, recognize their kin and distinguish them from other families and different species. They communicate and send carbon, the building block of life, not just to the mycorrhizas of their kin but to other members of the community. To help keep it whole. They appear to relate to their offspring as do mothers passing their best recipes to their daughters. Conveying their life energy, their wisdom, to carry life forward. The yews too were in this web, in relationship with their lifelong companions, and with people like me recovering from illness or just walking through their groves.

A few days after my final treatment, the paclitaxel doing its last work in my cells, Jean made the long trip over the Monashee to help me plant my garden, a celebration of getting back to the outdoors. "You look good, HH," she said, even though I was pale. We worked for hours, turning over the soil, the worms wiggling, grains moist, until our backs were sore, hands blistered, and we flopped down to drink kombuchas in the shade. The next day, we sowed beans, corn, and squash. When the seeds germinated, their radicals would signal the arbuscular mycorrhizal fungi, which would join the plants in an intimate web, as I imagined was happening among the yews, cedars, and maples across the lake. The tall waking cedars would be starting to infuse the sleepy little yews with sugars, which would use the energy to grow their shaggy bark and make drops of paclitaxel. As the maple leaves opened, they'd send sugary water to the cedars and yews in the shadows, helping them get enough to drink on the dry summer days. The yews might return the favors to the maples and cedars in late fall, sending sugary reserves from their green cells to help the neighbors slumber through the winter. Mycorrhizal fungi would begin to wrap around the mineral grains, waking up the mites and nematodes and bacteria.

I put a white seed into a hole I'd pressed in the ground. In a few weeks, the soil would be teeming, and by Mother's Day, life would awaken the seeds of the Three Sisters.

THE DAY I WAS DECLARED CANCER-FREE, Dr. Malpass warned that if the disease returned, I wouldn't survive. I wanted assurance I was all right, but he shrugged and said, "Suzanne, that is the mystery of life, and it is up to you to embrace it."

At home I sat under my maple, new leaves emerging, and listened to the squirrels climbing into her crown. She'd lost a large branch over the winter, and her sap was closing the wound, but she was still giving it her all to make leaves. She'd produced a bounty of new seeds, maybe her last, some that would produce young, others that the squirrels would eat.

The nagging question remained about the Mother Trees departing from the living. Would ailing mothers send their remaining carbon to their kin—a surge of delivering all she had—and would it move beyond the webbing of fungi enveloping their tiny roots and into their nascent leaves, helping them grow their budding photosynthetic tissues? Her last breath entering—becoming part of—her progeny.

I poked in my garden to check if my peas had germinated, and I was startled to find a maple seedling unfurling among the wavering tendrils.

15

PASSING THE WAND

Hannah slapped a mosquito the size of a B-52 bomber against her neck. As she stepped over the frayed plastic fringe encircling the roots of the Douglas-fir sapling, I said, "Touch her bark first, sweetie pie, to show your respect." She placed her hands on the young fir's smooth surface, then wrapped measuring tape around the trunk and called out the diameter—"Eight centimeters!"—the girth of a softball. Then she shouted "two"—code for "a wanting condition," yellowish needles a sign of root disease. Jean jotted the numbers on the data sheet. My niece Kelly Rose pointed the pocket-sized laser hypsometer at the roots and then at the terminal bud. "Seven meters tall," she called out. Nava and I were measuring a birch neighbor half the size of the fir, her base decorated with honey mushrooms.

We had returned to Adams Lake, one of the original sites where in 1993 I'd dug the grid of meter-deep trenches between fir and birch and wrapped plastic around the individual cylinders of roots to sever the mycorrhizal network that connected the trees. Twenty-one years later, in July 2014, we could see that the trees cut off from one another were suffering, their immune systems weak, their vitality bridled. Only thirty meters away was the thrifty control where I'd left the hyphal linkages intact.

It had been just over a year since I'd finished chemo, and Jean and I had brought Nava, Hannah, and Kelly Rose, fourteen, sixteen, and eighteen years old, to learn the ways of the forest and to see if the ecosystem really was a place where all were connected as one, the species wholly interdependent as my research had been showing me for decades, wisdom long held by Aboriginal peoples the world over. This was my chance to show all this to my girls while we spent a summer's day in the bush.

"Here, put on these bug nets," said Jean. She took green beekeeper hats from her work vest and showed the girls how to pull them over their twisted-up ponytails. "These are great," Kelly Rose said, her relief instant.

The site contained some of my oldest experiments. We finished measuring the fifty-nine trees in the trenched plot before moving into the untrenched control area, the understory lush with thimbleberry and huckleberry bushes. "At least it's cool under this birch," Nava said. She had shot up to five foot seven, as tall as Robyn, towering over Hannah and Kelly Rose, who'd settled in at the height of Grannie Winnie, five feet one and a half inches. All three girls had Grannie Winnie's quiet toughness—getting on with the job, not making much fuss, quick to laugh, kind and gentle, watching over one another. Not batting an eye at climbing a tree, swinging from a branch, grabbing the highest apple, landing on their feet, making an apple pie. Nava peeled a strand of the paper-thin bark and measured the tree's girth. "What made these?" she asked, pointing to tiny holes drilled in six perfect rows around the circumference.

"Sapsuckers," I said. "They peck into the bark to drink the sap and feed on the insects." Nava swerved as a real-live snitch vibrated toward her red vest, chirping *chu-chu-chu*. "Oh," I said, laughing, "hummingbirds like it too." The rufous jewel darted to a nest made with seed wings and spiderwebs, four tiny beaks stretching open. The next birch had been bent over by a moose who'd munched its tender shoots. On the banks of the Adams River, half a kilometer east, where the birches were thirty meters tall, the elk and deer and snowshoe hares ate the branches and buds too, and beavers built lodges with the waterproof stems, and grouse nested in the leaves, and sapsuckers and woodpeckers carved out cavities, later used by owls and hawks. The roots of these distinguished birches drank the water of the glacial-fed river, the water turning red with spawning salmon in the fall.

I'd been wondering if the birches were also nourished by the fish carcasses seeping back into the riverbanks.

Within a few hours, we discovered that the birch trees whose roots ran freely and connected with the firs were almost twice the size of those in the trenched plots, and they were free of disease. Compared to the birches we'd thinned alongside the creek nearby two decades ago, these were smaller, but they were healthy, the papery bark thick, lenticels (eyelets) compact, branches few, valuable for making baskets. The bigger birches were especially the kind that Mary Thomas, an elder of the Secwepemc Nation, said would be good for harvesting bark. Mary Thomas's grandmother Macrit showed her how to peel the bark so as not to hurt the tree, as her grandmother had shown her, and as Mary would show her own grandchildren. Teaching them how to leave the pulpy cambium intact so it would be primed to heal over, to ensure the tree seeded new generations. They used the bark to make baskets of all sizes, some for thimbleberries, cranberries, and strawberries. The impermeable bark of the bigger birches down by the river would be perfect for canoes, the luxuriant leaves for soap and shampoo, the sap for tonics and medicines, the best wood for bowls and toboggans. With care—planted in rich soil, with good neighbors, in proper numbers, and with roots unrestrained—even these upland birches could become prominent providers in the forest.

Woven among the birches, the firs were also a little bigger than where we'd trenched between them, and they were in prime condition. In the early years, the mycorrhizal connections with birch had helped the fir saplings grow taller, and in adulthood, this head start still mattered. Two decades later, firs performed better in the neighborhood of birches than where they'd been cut off from their neighbors or where they'd grown only among other firs. They had better nutrition—the rich birch leaves building the soil—and less *Armillaria* root disease, the bacteria along the birch roots providing a bundle of nitrogen and immunity with a potent mix of antibiotics and other inhibitory compounds. Grown intimately together, this forest had almost twice the productivity of the stands where we'd trenched between the species two decades earlier. This was the opposite of the usual foresters' expectations. They figured that fir roots free of birch interference would obtain more of the resource pie, as though the ecosystem worked as a zero-sum game—the adamant belief that greater total productivity cannot possibly emerge from species interactions.

Even more surprising to me was that birch benefited from fir too. Not only did birches likewise grow at twice the rate when intimately connected with firs than when alone, but they also had fewer root infections. The birches that had delivered food and good health to the firs when they were young were now being helped in reciprocity by the bigger firs as adults. Although the birches were retreating as the firs grew skyward, as happens naturally with the aging of these forests, their roots were still deep in the soil, their legacy of fungi and bacteria intact, lifeblood painted indelibly into the canvas. At the next major disturbance—a fire or an insect outbreak or a pathogenic infection— the roots and stumps would sprout again, bringing a new generation of birches, as much a part of the cycle as fir.

We sat under a sprawling birch for lunch. Salmon sandwiches we'd made at our campsite, and berries picked along the way, and cookies bought at the Vavenby General Store. Kelly Rose ate the blood-red thimbleberries one by one, as if selecting chocolates from a box. "Why are the plants so sweet under the birches, Auntie Suzie?" she asked.

Their roots and fungi draw water from deep in the soil, I told her, and with it bring calcium, magnesium, and other minerals, and this feeds the leaves so they can make sugars. The birches, with their cables of fungi, knit the other trees and plants together, and through their web share the nutritious soup drawn from the soil and also the sugars and proteins made by their leaves. "In the fall, when the birch leaves drop, they nourish the soil in return," I said.

Mary Thomas's mother and grandmother Macrit had taught her to show gratitude for the birches, to take no more than she needed, to place an offering in thanks. Mary Thomas had even called the birches Mother Trees—long before I had stumbled onto that notion. Mary's people had known this of the birches for thousands of years, from living in the forest—their precious home—and learning from all living things, respecting them as equal partners. The word "equal" is where Western philosophy stumbles. It maintains that we are superior, having dominion over all that is nature.

"Remember how I said the birches and firs talk to each other underground through a fungal web?" I asked the girls, putting my hand to my ear and my finger to my lips. The girls listened, their ears filled with mosquito songs. I told them I wasn't the first person to figure this out, that this was also the ancient wisdom of many Aboriginal people. The late Bruce "Subiyay" Miller of the Skokomish Nation, whose people live on the eastern Olympic Peninsula of Washington State, had told a story about the symbiotic nature and diversity of the forest, mentioning that under its floor "there is an intricate and vast system of roots and fungi that keeps the forest strong."

"This pancake mushroom is the fruit of the underground network," I said, handing an earthy bolete to Kelly Rose, who inspected its tiny pores and asked why it was taking so long for everyone to understand this.

I had been given a glimpse of these ideals—almost as a stroke of luck—through the rigid lens of western science. I'd been taught in the university to take apart the ecosystem, to reduce it into its parts, to study the trees and plants and soils in isolation, so that I could look at the forest *objectively*. This dissection, this control and categorization and cauterization, were supposed to bring clarity, credibility, and validation to any findings. When I followed these steps of taking the system apart to look at the pieces, I was able to publish my results, and I soon learned that it was almost impossible for a study of the diversity and connectivity of a whole ecosystem to get into print. *There's no control!* the reviewers cried at my early papers. Somehow with my Latin squares and factorial designs, my isotopes and mass spectrometers and scintillation counters, and my training to consider only sharp lines of statistically significant differences, I have come full circle to stumble onto some of the indigenous ideals: Diversity matters. And everything in the universe *is* connected—between the forests and prairies, the land and the water, the sky and the soil, the spirits and the living, the people and all other creatures.

We walked in the drizzle to where I'd planted conifers at different densities, to see how they liked growing in pure stands with few neighbors, or many. I knew every tree, every plot, every corner post. I knew where the larch was planted and the cedar. The fir and the birch. I showed the girls how this fir was planted too deep, that birch had been broken by a moose, this larch got pushed sideways by a black bear. I'd planted another spot every year for five years, but a tree would never take, and now it was a beautiful patch of lilies, what it was meant to be. In the mixed plots, cedars were luxuriant under birches, needing their cover to protect the pigments in their delicate leaves. When I stopped chattering and looked up, Jean and the girls were grinning.

We settled into measuring the Douglas firs planted at different densities. Without birch neighbors, up to 20 percent had become infected by *Armillaria* disease, more where the firs were tightly clustered. Their roots had grown into infection pockets in the soil, and the pathogens had spread under their bark, strangling the phloem, no birch roots to stop them. Some of the infected firs were still alive—needles yellowing—and others were long dead, their bark gray and flaking. In their place, other plants grew, and even some birches had seeded in, inviting the warblers and bears and squirrels. Some mortality wasn't a bad thing. It made room for diversity, regeneration, complexity. It kept the bugs down and created firebreaks. A lot of death, though, could cause a cascade of changes, rippling through the landscape, upsetting the balance.

Jean showed the girls how to start the bit of the increment corer in the bark of a fir. "If the corer doesn't take, don't try more than twice, so you don't injure her," Jean told them. Kelly Rose asked if she could try. Within minutes, she hit the pith—bull's-eye—and Jean inserted the core sample into a red straw, sealed the ends with masking tape, and labeled it.

In the high-density plots—where the planted firs were only a few meters apart—the understory was dark. The floor looked bare except for rusty needles, their acidity slowing down the cycling of nutrients. Gray branches snapped off as we worked our way among the trees. I imagined the mycorrhizal network had taken on the pattern of the plantings, wiring trees together as if they were rows of telephone poles. It would become a little more complex as the bigger trees spread their limbs and roots, taking over the growing space where others had died.

Our shins scraped, we moved on to a plot where the firs were spaced farther apart, up to five meters, their girth a little more robust. Seed had dispersed into the openings between the plantings over the years, some probably their kin, others the descendants of those removed, still others from firs in the surrounding forest. Fertilized by pollen of neighbors, or by firs in other valleys, ensuring the population was resilient. Some of these new trees were toddlers, others kindergartners, still others juniors, this patch of forest starting to look like a school-house, with diversity and kinship. The mycorrhizal network, I imag-ined, was becoming more complex as the forest aged, the biggest trees becoming the hubs—the Mother Trees. Eventually it would look like the web we'd mapped a few years back in the old-growth Douglas-fir forest.

After the last tree was measured, we followed a moose trail down to the river where we'd parked the truck. The forest was slowly tak-ing over my experiment, the replicates filled with surprises—a dozen tree species naturally seeding in from the timber edge, moose eating the planted birches, honey mushrooms infecting the trees, firs help-ing birches, young cedars huddling under the broadleaved trees for protection from the sun. This forest naturally knew how to rejuve-nate itself when allowed a proper start, seeding into soils that were receptive, killing my planted trees where they didn't belong, patiently waiting for me to hear what it was saying. *This data will be difficult to publish,* I thought to myself. Nature itself had blurred the rigidity of my experiment, my original hypotheses about species composition and density no longer testable due to the ingress of new trees. But I had learned so much more by listening instead of imposing my will and demanding answers.

As we drove the switchbacks over the mountain, the girls asleep in the back, Jean sorting the data sheets, I reflected on my good fortune with what the forest had shared with me over the course of so many years. In my first experiment testing whether birch transmitted car-bon to fir through mycorrhizas, I thought I'd be lucky to see anything, but then I detected a pulse strong enough to fuel the setting of seeds. I saw fir giving back to birch the energy it needed to build new leaves in the spring. And my posse of students confirmed the findings of reciprocity, not just between birch and fir but among all sorts of trees.

In making the mycorrhizal-network map, I thought we might see a few links.

Instead we found a tapestry.

With Yuan Yuan, I figured it would be a long shot if dying Douglas firs transmitted messages to ponderosa pines. But they did. Another of my students confirmed it in a second study, as did others in labs around the world. Then I considered it a gamble that Douglas-fir

Mother Trees would recognize their own kin, never mind that the signals might move through the mycorrhizal network—and *mon Dieu!* The firs recognized their relatives! The Mother Trees not only sent carbon to help support their mycorrhizal fungal symbionts, they somehow enhanced the health of their kin. And not only their kin, but of strangers too, and other species, promoting the diversity of the community. Was this all luck?

I think the trees had been telling me something all along.

I'd had a hunch those little yellow spruce seedlings back in 1980—the ones who'd sent me on this long journey of a lifetime—were suffering because their bare roots couldn't connect with the soil. Now I knew they lacked mycorrhizal fungi, whose hyphae would not only have extracted nutrients from the forest floor but also connected the seedlings to the Mother Trees, providing them with carbon and nitrogen until they could stand on their own. But their roots had been confined to their plugs, isolated from the old trees. The subalpine fir that had naturally regenerated on the outskirts of the Mother Trees, though, had been lush with sustenance.

But that lingering question since my illness still haunted me: If we are equal to everything in nature, do we share the same goals in death? To pass the wand as best we can. Passing onward to children the most crucial material. Unless the essential energy went *directly* to a Mother Tree's offspring, stem, needles, buds, and all—not just into the underground network—I couldn't be sure that the connection increased their fitness beyond that of the fungus.

Monika, a new doctoral student, had added another link in this chain of knowledge. In the fall of 2015, she started a greenhouse experiment with 180 pots. In each pot, she planted three seedlings: two kin and one stranger, with one of the kin seedlings designated as her "Mother Tree." The idea was that, once injured, the Mother Tree would have a choice of where to send the last of her energy: to her kin, the stranger, or into the earth. Monika grew the seedlings in mesh bags with variously sized pores to allow or inhibit mycorrhizal connections, and she injured some of the Mother-Tree seedlings with shears or western budworms. She then pulse labeled the Mother Trees with carbon-13 to trace where the carbon went.

As if to remind us of the capricious nature of nature, a heat wave knocked out the greenhouse's ceiling fans, killing part of the experiment. The greenhouse cat, a fat orange tabby, flicked his tail while Monika and I knelt near the rows, testing the dry-as-a-bone soil in pot after pot. Most of the seedlings were still alive. We were lucky. Even in greenhouse experiments, many environmental factors are under our control, but things still can go wrong. This pales in comparison to the myriad calamities that can happen in even the most well-conceived field experiment, especially over the decades it takes to examine long-term patterns. *No wonder most scientists conduct their research in a lab,* I thought to myself.

But we didn't ditch the experiment. Besides, Monika's kin seedlings were many times the size of Amanda's, and I was burning to know whether they were robust enough sinks to draw the carbon released by the injured Mother Trees into their shoots. Using the survivors, the day came when Monika and I were scrolling through graphs of data as if watching a movie. All of the factors we tested were significant—whether the seedlings were related to the Mother Trees, whether they were connected, or whether they were injured.

Monika's Mother-Tree seedlings transmitted more carbon to kin than strangers, as Brian and Amanda had found. But unlike the earlier study, where we'd only detected carbon moving into the mycorrhizal fungi of the kin seedlings, Monika now found that *it went straight into their long leaders.* The Mother-Tree seedlings flooded the mycorrhizal network with their carbon energy, and it advanced into the needles of her kin, her sustenance soon within them. *Et voilà!* The data also showed that injury, whether by western spruce budworm or the shears, induced the Mother-Tree seedlings to transfer *even more carbon* to her kin. Facing an uncertain future, she was passing her life force straight to her offspring, helping them to prepare for changes ahead.

Dying enabled the living; the aged fueled their young.

I imagined the flow of energy from the Mother Trees as powerful as the ocean tide, as strong as the sun's rays, as irrepressible as the wind in the mountains, as unstoppable as a mother protecting her child. I knew that power in myself even before I'd uncovered these forest conversations. I'd felt it in the energy of the maple in my yard, flowing into me as I contemplated Dr. Malpass's wisdom about embracing the mystery of life, sensing that magical, emergent phenomena when we work together, the synergy that reductionist science so often misses, leading us to mistakenly simplify our societies and ecosystems.

The trees of the next generation with genes most adaptable to change—whose parents have been shaped by a variety of climatic conditions, those attuned to the stresses of their parent, with robust defense arsenals and shots of energy—ought to be the most successful in rebounding from whatever tumult lies ahead. The practical application—what this might mean for forest management—is that elders that survived climate changes in the past ought to be kept around because they can spread their seed into the disturbed areas and pass their genes and energy and resilience into the future. Not only a few elders, but a range of species, of many genotypes, kin and strangers, a natural mix to ensure the forest is varied and adaptive.

My wish is that we might think twice about salvage harvesting the dying Mother Trees, might be compelled to leave a portion behind to take care of the young, not merely their own but those of their neighbors too. In the wake of diebacks from droughts, beetles, budworms, and fires, the timber industry has been cutting vast swaths of forest, the clear-cuts coalescing over whole watersheds, entire valleys mowed down. The dead trees have been considered a fire risk, but more likely a convenient commodity. Great numbers of healthy neighbors have also been captured for the mills as collateral damage. This salvage clear-cutting has been amplifying carbon emissions, changing the sea-

Giving a TED walk in Stanley Park at TED Vancouver, 2017

sonal hydrology in watersheds and in some cases causing streams to flood their banks. With few trees left, the sediments are flowing down rivulets and into rivers already warming with climate change, harming the salmon runs even further.

This brings me to another adventure, one I'm still exploring because it speaks so graphically to the species connections we overlook. Scientists before me have discovered that the nitrogen from decayed salmon lives in the rings of trees along the rivers from where they came. I wanted to know whether salmon nitrogen was absorbed by mycorrhizal fungi of the Mother Trees and transmitted through their networks to other trees deeper in the forest. Even more, were the salmon nutrients in the trees declining with the reduction in salmon populations and habitat loss, causing the forests to suffer? If so, could this be remedied?

MONTHS AFTER MONIKA'S EXPERIMENTS, I was at Bella Bella on the midcoast of British Columbia, in the salmon forests of the Heiltsuk people. Our skiff glided into a pristine inlet, and our Heiltsuk guide, Ron, pointed to ochre pictographs marking a clan territory. Silken Pacific mist poured down the vertical rock wall and over the monumental trees. With me were Allen Larocque, my new doctoral student who would investigate the patterns of the fungal networks, and postdoctoral fellow Dr. Teresa "Sm'hayetsk" Ryan of the Tsimshian Nation, the people of the Skeena River to the north. Teresa was a traditional cedar basket weaver as well as a salmon-fisheries scientist on the Canada-U.S. Pacific Salmon Commission, Joint Chinook Technical Committee, among the many hats she wore. She wanted to know, as an Aboriginal person and a scientist, whether restoring the traditional fishing practices using tidal stone-trap technology could reinvigorate the salmon populations, perhaps to levels seen before the colonists took control of the fishery. This in turn might nourish the cedars from which she gathered bark.

We were in search of the bones of salmon carried into the forest by bears and wolves and eagles. The bones were all that were left once the flesh was eaten and the residual tissue decayed, nutrients seeping into the forest floor. In this inlet, Dr. Tom Reimchen of the University of Victoria and Dr. John Reynolds of Simon Fraser University had discovered salmon nitrogen in rings of cedars and Sitka spruces, and in the plants, insects, and soils. Allen would start our study of how mycorrhizal fungi might transmit the salmon into the trees, and possibly between trees, by determining how the mycorrhizal fungal communities differed alongside streams with various salmon population sizes. Could a difference in the fungi, in their ability to transmit the salmon nutrients, help account for the great fertility of these rain forests? I could barely contain my excitement as Allen, Teresa, and I jumped into the sedges with our hip waders and headed to shore.

"Bear path," said Teresa, pointing at a trail. "They've been here recently."

"Let's keep going." I was like a dog pulling on its leash.

We easily followed the trail into the wall of salmonberry along the shore, where the prickly canes stood thick. After half an hour of crawling on hands and knees in the humus, Teresa suddenly said, "You guys are nuts. You're asking for trouble with these fresh bear signs." She headed back to the boat to wait with Ron.

I looked at Allen to gauge his comfort level, and he didn't seem nervous. "If I were a bear, I'd take my salmon to where I wouldn't be disturbed," I said, thrilled he was up for the adventure. We kept crawling along, through a tunnel carved in the salmonberries, toward a fifty-meter-tall cedar on a high bench, her leader forked in a candelabrum, what the Heiltsuk called a Grandmother Tree.

Each bear preying on the spawning salmon transported some 150 fish per day into the forest, where the roots of the trees foraged for the decaying protein and nutrients, the salmon flesh providing more than three-quarters of the tree's nitrogen needs. The nitrogen in tree rings derived from salmon was distinguishable from the soil's nitrogen because fish at sea get enriched with the heavy isotope nitrogen-15, which serves as a natural tracer of salmon abundance in the wood. Scientists could use the year-by-year variation in tree-ring nitrogen to find correlations between salmon populations and changing climate, deforestation, and shifting fisheries practices. An old cedar tree could hold a thousand-year record of salmon runs.

I shouted *Yoohoo!* as we approached the ledge of the Grandmother cedar, despite my call being muffled by a wall of salmonberry leaves. A

grizzly out here would mean a quick death. Still, I felt peaceful. After
chemo, this was bliss. And I was much calmer than I'd recently been
on the big TED stage in Banff, where cameras and a thousand people
had tracked my every move. I'd stepped into the bright lights thanking my lucky stars that Mary had made me wear a black coat over my
blue shirt, an old favorite, because she'd spotted its missing button. I
delivered my talk as though the audience were a sea of nodding cabbages. *I did it,* I thought as I walked off the stage, pride in overcoming
my shyness, to speak from my heart, to unfurl what I've learned so
that people could take what they needed, flooding me. "I've always
known this about trees deep down inside of me," a woman wrote
from Chicago after seeing the video. Robert Krulwich of *Radiolab*
contacted me to create a podcast. *National Geographic* wanted to write
an article and make a film. I received thousands of emails and letters.
Kids, mothers, fathers, artists, lawyers, shamans, composers, students.
People from every corner of the world expressing their own connections with trees through their stories, poems, paintings, films, books,
music, dances, symphonies, festivals. "We'd like to design our city in
a way that mimics the patterns of mycorrhizal connection," wrote a
city planner from Vancouver. The concept of the Mother Tree and her
connections to those around her had even made it into Hollywood,
as a central concept to the tree in the film *Avatar.* How the film resonated with people reminded me how naturally crucial it is for people
to connect to mothers, fathers, children, family—our own and the
families of others—and to trees and animals and all of the creatures
of nature, as one.

I'd taken my message and gone out with it, and a bracing surge of
responses came in return. People cared about the forest and wanted
to help.

"What we're doing isn't working," a government forester wrote.
Music to my ears. We discussed how to leave Mother Trees to help
heal the land following a harvest. Not enough foresters have embraced
this yet, but at least there is a small beginning.

Allen and I crept up and peered along the bench. *"Putain de merde!"*
I shouted. "Look!" Under the boughs of the old Mother Tree was a
cozy, mossy bed large enough for a mama bear and her cub. Dozens
of white salmon skeletons gleamed from the carpet, the flesh long

decayed, the vertebrae unhinged, the fine corsets of bones folded like butterfly wings, the scales and gills asunder, the essence of the fish slowly absorbed by the roots, transmitted into the wood of the tree, passed to the next life.

Tree bones.

Allen and I collected soil from under the bones and, for comparison, from places where there were no bones. We returned to Teresa and Ron, jumping onto the boat from the high-tide line and storing the samples on ice to prevent the degradation of the microbial DNA. Ron puttered away from the shore and skimmed over the stone wall that followed the contour of the shoreline, from one edge of the estuary to the next. The wall was one of hundreds of tidal traps built along the Pacific coastline by the Heiltsuk people, similar to those built by the Nuu-Chah-Nulth, Kwakwaka'wakw, Tsimshian, Haida, and Tlingit Nations—to harvest salmon passively, keep track of the populations, and adjust harvests accordingly. They collected the fish trapped at low tide, releasing the biggest egg-bearing females to continue up the river to spawn. They smoked, dried, or cooked the fish, buried the guts in the forest floor, and returned the bones to the waters to nourish the ecosystem. This practice enhanced the salmon populations and the productivity of the forests, rivers, and estuaries. The forests, rich with salmon, returned the favor by shading the rivers, shedding nutrients into the waters, and providing habitat for the bears, wolves, and eagles.

Teresa explained that when the colonists took jurisdiction of the waters and forests, they forbade use of the stone traps. The salmon were overfished within the first two decades and have yet to recover fully. Climate change and a warming Pacific Ocean have created new problems by exhausting the fish on their marathon from the ocean, reducing their success at reaching the natal spawning streams. It's part of a general pattern of destroying interconnecting habitats. To the north on Haida Gwaii, the last of the cedars, some more than a thousand years old, are being clear-cut on Graham Island, leaving the forest along the spawning rivers degraded and the Haida wondering what will happen to their way of life.

When will this stop, this unraveling?

As we sped out of the inlet toward Bella Bella, Ron pointed starboard to a humpback surfacing a few hundred meters away. From out of nowhere, dozens of white-sided dolphins joined our boat, arcing over the water, somersaulting and whistling to one another. I was so astonished, so uplifted, that I stood, Allen and Teresa too, as the salt water splashed over us.

This study is ongoing, but our early data show that the mycorrhizal fungal community in the salmon forest differs depending on the number of salmon returning to their natal streams. We still don't know how far into the forest the mycorrhizal network is transporting the salmon nitrogen, and if—or how—restoration of the tidal stone traps might affect forest health, but we are starting new research and reconstructing some of the stone walls to find answers. I've been wondering too if we should check whether salmon also nourish the mainland forests from rivers that run inland. Do spawning salmon feed the cedars and birches and spruces along the rivers that run thousands of kilometers into the mountains? Such as along the Adams River running below my experiment. Salmon in this way connecting the ocean with the continent. The Secwepemc people knew how vital salmon was to the interior forests, and to their livelihoods, and they'd cared for the populations according to far-reaching principles of interconnectedness.

THANKSGIVING THAT YEAR found me driving home past clearcuts as chain saws were bringing down the beetle-infested Mother Trees before their seeds had germinated in the turned-up duff. Slash piles of elders stood as tall as apartment buildings, access roads crisscrossed the valleys, and creeks were clogged with sediment. Planted seedlings stood encased in white plastic tubes like crosses.

The cracks are in plain view.

I come from a family of loggers, and I am not unmindful that we need trees for our livelihoods. But my salmon trip showed that with taking something comes the obligation to give back. Of late I've become increasingly enchanted by the story told by Subiyay, who talks of the trees as *people*. Not only with a sort of intelligence—akin to us humans—or even a spiritual quality perhaps not unlike ours.

Not merely as equivalent to people, with the same bearings.

They *are* people.

The Tree People.

I don't presume to grasp Aboriginal knowledge fully. It comes from a way of knowing the earth—an epistemology—different from that of my own culture. It speaks of being attuned to the blooming of the bitterroot, the running of the salmon, the cycles of the moon. Of knowing that we are tied to the land—the trees and animals and soil and water—and to one another, and that we have a responsibility to care for these connections and resources, ensuring the sustainability of these ecosystems for future generations and to honor those who came before. Of treading lightly, taking only what gifts we need, and giving back. Of showing humility toward and tolerance for all we are connected to in this circle of life. But what my years in the forestry profession have also shown me is that too many decision-makers dismiss this way of viewing nature and rely only on select parts of science. The impact has become too devastating to ignore. We can compare the condition of the land where it has been torn apart, each resource treated in isolation from the rest, to where it has been cared for according to the Secwepemc principal of *kwseltktnews* (translated as "we are all related") or the Salish concept of *nôcaʔmat ct* ("we are one").

We must heed the answers we're being given.

I believe this kind of transformative thinking is what will save us. It is a philosophy of treating the world's creatures, its gifts, as of equal importance to us. This begins by recognizing that trees and plants have agency. They perceive, relate, and communicate; they exercise various behaviors. They cooperate, make decisions, learn, and remember—qualities we normally ascribe to sentience, wisdom, intelligence. By noting how trees, animals, and even fungi—any and all nonhuman species—have this agency, we can acknowledge that they deserve as much regard as we accord ourselves. We can continue pushing our earth out of balance, with greenhouse gases accelerating each year, or we can regain balance by acknowledging that if we harm one species, one forest, one lake, this ripples through the entire complex web. Mistreatment of one species is mistreatment of all.

The rest of the planet has been waiting patiently for us to figure that out.

Making this transformation requires that humans reconnect with

Hannah, age twenty-one, working in
the bush and eating huckleberries,
July 2019

nature—the forests, the prairie, the oceans—instead of treating everything and everyone as objects for exploitation. It means expanding
our modern ways, our epistemology and scientific methodologies,
so that they complement, build on, and align with Aboriginal roots.
Mowing down the forests and harvesting the waters to fulfill our wildest dreams of material wealth *just because we can* has caught up to us.

I crossed the Columbia River at Castlegar, only half an hour from
home, anxious to see Hannah and Nava, grateful that Mary had made
the trip north for the Canadian Thanksgiving. The river was low, the
natural flow controlled by the Mica, Revelstoke, and Hugh Keen-
leyside Dams upstream—three of the sixty in the Columbia watershed. These dams meant the loss of salmon from the Arrow Lakes
and the flooding of the villages, burial grounds, and trade routes of
the Sinixt Nation, whose ancestral territory spans from the Monashee
Mountains east to the Purcells and from the Columbia headwaters
to Washington State. I wondered what this land looked like before
the Canadian government declared the Sinixt Nation extinct, then
dammed, clear-cut, and mined their landscape. The Sinixt people are
resilient, though, continuing to uphold *whuplak'n*—the law of the
land—joining together to help restore the Columbia watershed.

I arrived home, the moon high above snow-dusted mountains, where Mary and the whole family had converged. This Thanksgiving turned out particularly memorable because the tea-scented candles on the table tipped over, and flames licked around the turkey. I looked up from stirring the gravy as Don—his new girlfriend was off with her own kids—threw the pot of water for the brussels sprouts over the burning bird, and Robyn and Bill doused the napkins with their glasses of wine. Grannie Junebug carried her trifle past Oliver reading a Harry Potter book on the floor.

Family. In all its imperfection, and stumbles, and small fires. We were there for one another when it counted.

In spite of the clear-cuts, and my worries about work and climate change, and my health and my children, and everything else, including my precious trees, it was great, simply great, to be home, all of us together.

HANNAH FOLLOWED ME into the grove of hemlocks among the rockpiles below the black hole in the cliff—a portal leading to kilometers of tunnels blasted into the mountain a century ago by miners in search of copper and zinc. We dug a soil pit among the trees, some of the mineral grains green, others rust, our hands protected by surgical gloves, arms covered by long sleeves. Seepage from the portals was loaded with copper, lead, and other metals, and these had contaminated the forest floor. The metals combined with sulfides in the ore with the help of bacteria to form acid-rock drainage, which had leached from waste-rock piles deep into the soil. And yet trees were growing here, even if slowly, giving their all to fuel the recovery of the forest.

It was the summer of 2017. We were at the Britannia Mine—forty-five kilometers north of Vancouver on the shores of Howe Sound on the unceded territory of the Squamish Nation—the largest mine in the British Empire, opened in 1904 to extract the ore bodies that had formed when volcanic pyroclast flowed onto sedimentary rock and the metamorphosed result came into contact with plutonic intrusions. The miners had quarried the faults and fractures where the rich ore lay, boring right through Britannia Mountain, from Britannia Creek on the northern flank to Furry Creek on the southern side, covering an area of about forty square kilometers. They left behind two dozen portals to 210 kilometers of tunnels and shafts, which stretched from 650 meters below sea level to 1,100 meters above.

The men had transported the ore from inside the mountain along rails that emerged into daylight at the portals, where they loaded it onto rail carts and tramways, leaving the waste rock in piles. Even after the mine closed in 1974, it remained one of the largest point sources of metal pollution to the marine environment in North America. Tailings and waste rock were used to fill in the shoreline, and Britannia Creek, containing kilograms of copper, flowed clear but devoid of life into Howe Sound, killing ocean life for at least two kilometers along the shore. Britannia Creek's water was so toxic at the mine's closing that Chinook salmon fry, when introduced, died within forty-eight hours. With years of remediation, salmon have returned to spawn successfully in Britannia Creek, and the shoreline at Britannia Beach is alive again, with plants and invertebrates on the rocks and dolphins and orcas in Howe Sound.

These are signs that the earth can be forgiving.

I'd come here with Hannah at the request of Trish Miller, an environmental toxicologist, to assess the impact on the surrounding forest from the waste-rock piles. The effects were not isolated to the creeks but had reached farther into the forest, and she wanted a broader assessment than is usually done. I had jumped at the chance to work with Trish, having listened to her talk of environmental remediation for many years as friends when our kids were small. I was curious to discover the capacity of the forest to heal a broken ecosystem, of the old trees to seed into the raw earth, of the fungal and microbial networks to mend the damage. How well were the trees growing in the halos of metal-contaminated forests around the waste-rock piles? Was the forest recovering? Should we do more, or could the forest slowly heal on its own?

How profound a wound could the forest suffer before healing became impossible?

Hannah and I found the portals hidden among the hemlocks, the shawls of the trees enveloping the yawning gateways to the caverns. Alders and birches lined the hand-carved mine roads and rail tracks that led from the tunnels high in the crags to the separator mill at the shoreline below. Mosses and lichens covered the camps where the miners had slept, and the town sites where their families had lived were silent. The humus in the halo forest around the waste-rock piles was more barren than in the surrounding uncontaminated forests, but the roots of the trees had entwined the exposed stones, and a smattering of acid-loving false azaleas and black huckleberries and bracken ferns had found a foothold. As we stood under the hemlock branches dripping with rain, I felt that if there was anywhere that the earth held the power to heal, it would be here on the Pacific Coast in one of the most productive rain forests in the world.

This was also a chance to show Hannah how to assess the disruption—to the trees and plants, soils, and mosses—and the capacity for nature to recover, even when her veins had been bled at the surface. These waste-rock piles were smaller in comparison to the hundreds of meters of a clear-cut, a thousand for the clear-cuts coalescing across valleys, and thousands in the open-pit copper mines around the world. The disturbance by clear-cutting is acute, but the forest can readily recover where its floor is left intact, whereas removing the soil and mining metals from deep inside the earth has a chronic effect on the forests and streams.

"It's good the trees are coming back," Hannah said, coring a small western hemlock. It was one of dozens—lined up like foot soldiers— that had found a niche in the decaying wood. Their seed had dispersed in from adjacent healthy forests, their roots finding purchase in rotting nurse logs where fungal symbionts absorbed scarce nutrients, spongy cellulose sopped up water, and light shafted in thinly from the overstory. Hannah's tree was growing at only half the rate of the nearby elders—her roots shallower, crown sparser—but I knew she would make it. My master's student Gabriel had found that even hemlock saplings like these, whose roots gripped old nurse logs, could also connect with nearby Mother Trees, and they received carbon from the powerful crowns until they themselves were self-sufficient providers. The plant community in this understory was recovering too, with half of the old-growth shrubs and herbs now present in small patches, most of them acid lovers like the hemlock, slowly changing the soil and speeding the cycle of nutrients. These feedbacks were crucial in helping the trees regain their momentum. In the soil pit, I measured the depth of the forest floor—the litter, fermentation, and humus layers—and it was already about half that of the adjacent healthy areas.

When I peeled back the forest floor to look at the underlying mineral soil, a bronze centipede, big as a salamander, writhed onto my hand. "Ah!" I yelled, throwing the arthropod against a log, where it tumbled into some humus. The centipede was furious, wriggling so fast that the dirt churned. A sign—a stunning one—that the forest floor was in recovery. It burrowed out of sight to continue its day's work, eating the smaller bugs, them eating even smaller ones, and in eating and excreting, cycling the nutrients, a chain of actions to help the trees grow. Hannah and I ate our chocolate chip cookies before measuring and recording the depth and texture of the soil, the height and age of the trees, the species and cover of the plants, signs of birds and animals.

We drove five kilometers farther up the mountain and surveyed the plants and soils across a scree slope of waste rock, the 70 percent angle so steep that a rope had been strung to help workers rappel down. The talus was mostly bare in the middle, some lichens creeping over the rock shards, the odd grass rooted. The hemlock germinants that had found a grain of humus to root in were sickly pale—chlorotic—from insufficient nitrogen, reminding me of the little yellow seedlings in the Lillooet Mountains so long ago. Hannah kept pace behind me as we poked across the steep talus. The hemlocks seeded in from surrounding Mother Trees were increasingly robust as we approached the timberline. At the forest edge, shrouded in the mist, the saplings were bigger, their foliage brighter, mycorrhizas entwined with the minerals, building soil themselves. Bit by bit, with the help of the Mother Trees, the creatures—fungi and bacteria, plants and centipedes—were working together to heal the wounds of this exploited, majestic place.

"Bringing in soil from the old forest would also help," I said, recalling how Grannie Winnie built her garden with compost, burying at the base of the raspberry canes the guts of the fish that Grampa Bert caught, much as the Heiltsuk and the bears and wolves nurtured the Grandmother cedars with the bones of the salmon, giving back, completing cycles. I swear the berries were sweetest where Grannie worked. I loved that Hannah was following me, just as I'd accompanied Grannie through her patches of corn and potatoes.

"You could plant birches and alders here too," said Hannah, suggesting we collect seed from the alder along the stream and from the birches along the old mining road.

"Good thinking," I said, "and in clusters, not in rows." Trees need to be near one another, to establish in receptive soil, to join together to build the ecosystem, mix with other species, relate in patterns that produce a wood-wide web, because the forest becomes resilient from this complexity. Scientists now are more willing to say that forests are complex adaptive systems, comprised of many species that adjust and learn, that include legacies such as old trees and seed banks and logs, and these parts interact in intricate dynamic networks, with information feedbacks and self-organization. Systems-level properties emerge from this that add up to more than the sum of the parts. The properties of an ecosystem breathe with health, productivity, beauty, spirit. Clean air, clean water, fertile soil. The forest is wired for healing in this way, and we can help if we follow her lead.

We reached the mound of waste rock at the top portal, the blasts having exposed a cavernous scar a few hundred meters high and just as wide, the waste rock in knuckled piles at the base. The air was thinner, the clouds billowing over the granite turrets, cold rain pelting us. The mountain hemlocks around the portal were still exuberant, their needles like velvet, branches ragged from the wind, tops curved by the tonnage of snow. Their roots spread under the forest floor like veins across old hands, cycling the granite into wood, feeding the plants and animals.

But then, at the scar, where the rock glinted with deep-earth metals, the roots stopped. Like the rails that stopped midair at the portal below, as though men had been hurtled into the river to their deaths. These gouges were too profound for the roots to continue, the unearthed rock too raw to offer nourishment, the water too acid to drink, the wounds impossible to knit. The metallic rock glistened under water seeping from the crags, the lichens and mosses still absent from the grains even after a century of peace. I could see Hannah's shock that the earth sometimes simply cannot bear—cannot recover from—too enormous an injury. There is only so much hurt it can

Mother Tree in the inland rain forest near Nelson, British Columbia

take. Some connections are too broken, the blood too drained, even for the magnificent, healing roots and tenacity of a powerful Mother Tree.

We descended to the lowest portal. The gash creating a mine at this elevation was smaller—the forest here would recover. Hannah counted the rings from our day's final coring and wrote down "eighty-seven years." She inserted the pencil of rings back into the tree, sealed the wound with pitch, and patted the bark.

"The beauty," I said, "is that with a little momentum, a little help at this site, the plants and animals will come back." They'd make the forest whole again, help it recover. The land wanted to heal itself. Just as my body did, I thought, grateful to be here, continuing my work, teaching my daughter. Once the system hits a tipping point, once good decisions are made and acted upon, and when parts and processes are enmeshed again, and the soil rebuilt, recovery is possible— at least in some places. We gathered our gear to wind down the slope, the soil still speckled coppery green, the seeping water still a little acidic, but all of it slowly changing.

Carpets of lush seedlings swished around our ankles. Columns of taller hemlocks marched down fallen logs, their leaders lusty in their search for the sun, roots entwined with the wood. "I think I want to be a forest ecologist, Mama," my daughter said, running her hands along the saplings' feathery needles.

I stopped and looked back. Silhouetted in the setting sun, rising above the others, rooted in the volcanic rocks that nourished her, was the Mother Tree of this wide swath of seedlings. Her limbs were spread like arms, gnarled from centuries of snow, scars long healed over, fingertips loaded with cones. I was calm, happy, but also in need of rest. A classroom in Virginia had sent me a poem entitled "Mama Tree," in which a Mother says to us all: *Goodnight, my loves; it's time to sleep.* This evening I would take the little trail down to the Squamish River and sit on the riverbank with the herons and close my eyes in the warm air.

Hannah took the camera and GPS unit from her vest pocket to snap a photo and log the location of the old Mother Tree and her brood of seedlings. "We can put this in our report," she said, her ability to *see* the forest growing boundlessly.

With the sun sinking behind the sprawling crown of the Mother Tree, a bald eagle landed on her highest branch, scattering her cones. He angled his white head to stare straight down at us. I exhaled sharply, my breath joining a rush of mountain air. I like to think it was carried up to the eagle, because right then he ruffled his prodigious wings. *Now I know why.* I know why these seedlings are healthy in spite of the damage and ravages, unlike the little yellow seedlings from so long ago in the Lillooet Mountains, the ones that received the promise of how I would dedicate my life. The seeds here had germinated in the vast mycorrhizal network of this parent.

Their nascent roots drank from the nutritious soup supplied through her web. The shoots received messages about her past struggles, giving them a head start.

Their response was this plumage of emerald.

The eagle suddenly lifted, caught an updraft, and vanished past the peaks. There is no moment too small in the world. Nothing should be lost. Everything has a purpose, and everything is in need of care. This is my creed. Let us embrace it. We can watch it rise. Just like that, at any time—all the time—wealth and grace will soar.

Hannah stuffed the soil samples in her pack. Ferns shuddered in the raindrops, and she pulled on her hood. She peeked out to see where the eagle had flown, then pointed to it joining a companion over the granite arêtes.

The wind whipped through the needles of the Mother Tree, but she stood steadfast. She had seen nature in countless forms: hot summer days when the mosquitoes swarmed; rain that came in sheets for weeks; snow so heavy some of her branches snapped; periods of drought followed by long damp spells. The sky turned scarlet, her limbs on fire, blood rising to a battle cry. She would be here for hundreds more years, guiding the recovery, giving it her all, long after I was gone. *Farewell, dear Mama.* Tired, I fumbled to do up my vest. Hannah slung her heavy pack over her shoulders, adjusting the load and cinching the buckles, barely noticing the weight.

She took my shovel to lighten my burden and gripped my hand to lead us back home.

THE MOTHER TREE PROJECT

I began the Mother Tree Project in 2015, during my rebirth after cancer. It is the biggest experiment I've ever conducted, with a guiding principle of retaining Mother Trees and maintaining connections within forests to keep them regenerative, especially as the climate changes.

The Mother Tree Project consists of nine experimental forests located across a "climate rainbow" in British Columbia—from hot and dry forests in the southeast corner of the province to cold and wet stands in the north-central interior. We are examining the structures and functions of the forests—how webs of relationships play out in real environments and change with forest-cutting patterns that retain various numbers of Mother Trees and plantations that contain different tree-species mixtures. We want to make educated guesses about which combinations of harvesting and planting will be most resilient to the stresses our planet is facing, how the healthiest connections can thrive alongside our needs to use resources from the forest.

Our goal is to further develop an emergent philosophy: complexity science. Based on embracing collaboration in addition to competition—indeed, working with all of the multifarious interactions that make up the forest—complexity science can transform forestry practices into what is adaptive and holistic and away from what has been overly authoritarian and simplistic.

By now, everyone knows about the consequences of climate change, and almost no one has escaped its direct wrath. Concentrations of carbon dioxide have exploded from a level of 285 parts per million (meaning 285 molecules of carbon dioxide in 1 million molecules of air) in 1850 to 315 parts per million in 1958. As I sit here writing, concentrations have exceeded 412 ppm, and at the rate we are going, when Hannah and Nava might be raising children, we will reach the 450 ppm level that scientists consider a tipping point.

But I am hopeful. Sometimes when it seems nothing will budge, there's a shift. Based on my research, the free-to-grow policy was revamped back in 2000 to tolerate a few birches and aspens in certain regions of the province, though fundamental attitudes had not fully changed—these leafy trees were still viewed as competitors, irritants. But now there are young foresters out there on the land, writing thoughtful prescriptions and applying the ideas of saving old trees and encouraging forest diversity.

We have the power to shift course. It's our disconnectedness—and lost understanding about the amazing capacities of nature—that's driving a lot of our despair, and plants in particular are objects of our abuse. By understanding their sentient qualities, our empathy and love for trees, plants, and forests will naturally deepen and find innovative solutions. Turning to the intelligence of *nature itself* is the key.

It's up to each and every one of us. Connect with plants you can call your own. If you're in a city, set a pot on your balcony. If you have a yard, start a garden or join a community plot. Here's a simple and profound action you can take right now: Go find a tree—*your tree*. Imagine linking into her network, connecting to other trees nearby. Open your senses.

If you want to do more, I invite you into the heart of the Mother Tree Project to learn techniques and solutions that will protect and enhance biodiversity, carbon storage, and myriad ecological goods and services that underpin our life-support systems. Opportunities are as endless as our imagination. Scientists, students, and the general public who want to take part in this interdisciplinary research deep in the forest and be part of a citizen-science initiative, a movement to save the forests of the world, can find out more at http://mothertree project.org.

Vive la forêt!

ACKNOWLEDGMENTS

It's nigh on impossible to acknowledge fully all the many individuals whose support and devotion helped me carry out the work detailed in *Finding the Mother Tree*. Each chapter reflects a community effort, and I am forever indebted to all of those who lived, worked, and learned alongside me to create this story and bring it to light. The loving contributions of my family and friends, my students, teachers, and colleagues, and my writing coach, agent, and publishers provided me with the strength, endurance, and courage to persevere.

I owe the initiation of this book to Doug Abrams and Lara Love Hardin at Idea Architects. Without their interest, insight, and creativity, this book would be much less than it is. I am especially grateful for my close work with my writing coach, Katherine Vaz. Katherine tapped my memories and ideas for every chapter, coaxed out important details, eliminated awkward passages, and folded the stories together in a way that made a person want to keep reading. She supported and encouraged me from the first words until the end, and by the time we were finished, I felt she knew as much about my life as I did. Our friendship has grown from the moment we met. To brilliant Katherine go my profound thanks for helping make this book radiant.

I am grateful to my editor at Knopf Doubleday Publishing Group, Vicky Wilson. I thank her for taking an interest in a manuscript on trees, and for knowing that the same worldviews that were degrading our forests were also convulsing our society, and that fixing these problems required a deep look at ourselves, our place in nature, and at what nature has to teach us. My whole family and I are indebted to Vicky for her idea to include historic photographs throughout. To Vicky, thank you for seeing the value of this book and bringing it to vivid life.

My editor at Penguin in the UK, Laura Stickney, helped sharpen the scientific passages with a careful eye. Thank you, Laura, for your attention and crucial editorial skills in the book's final stages.

To my family, this book is my love letter to you. It is a poem of gratitude to my maternal grandparents, Winnie and Bert, and the extended Gardner and Ferguson families, and my paternal grandparents, Henry and Martha, and the Simard and Antilla families, who taught me about the water and streams and forest. From them I learned how we were settlers on this land and how to live with joy in spite of hardship. And most of all, to my parents, Ellen June Simard and Ernest Charles (Peter) Simard, and my siblings, Robyn Elizabeth Simard and Kelly Charles Simard; every line is about us, the place we came from, and the forests that shaped us. This book is also a gift to their families, especially Oliver Raven James Heath, Kelly Rose Elizabeth Heath, Mathew Kelly Charles Simard, and Tiffany Simard, in whose lives these stories live on.

To my beautiful friends, I love your quirks and quarks as much as you love mine. Especially, thank you Winnifred Jean Roach (née Mather) for being the most beautiful best friend a person could ever hope for, and with whom I have dedicated my life to the forest over the past four decades. Thank you, too, Barb Zimonick, for being my technician at the Forest Service for more than a decade, for looking after the accounts and trucks and equipment and summer students, even as we worked long hours out of town when your kids were little. To Barb and her family for all of those years go my heartfelt appreciation.

It is an impossible task to acknowledge fully all of the students, postdocs, and research associates at the University of British Columbia who have helped and inspired me to carry out this research. Your work is embedded in the science I speak of in these chapters, and I thank you in the order that you have studied with me: Rhonda DeLong, Karen Baleshta, Leanne Philip, Brendan Twieg, François Teste, Jason Barker, Marcus Bingham, Marty Kranabetter, Julia Dordel, Julie Deslippe, Kevin Beiler, Federico Osorio, Shannon Guichon, Trevor Blenner-Hassett, Julia Chandler, Julia Amerongen Maddison, Amanda Asay, Monika Gorzelak, Gregory Pec, Gabriel Orrego, Huamani Orrego, Anthony Leung, Amanda Mathys, Camille Defrenne, Dixi Modi, Katie McMahen, Allen Larocque, Eva Snyder, Alexia Constantinou, and Joseph Cooper. To my postdocs and research associates, you are the unsung heroes of this work:

Drs. Teresa Ryan, Brian Pickles, Yuan Yuan Song, Olga Kazantseva, Sybille Haeussler, Justine Karst, and Toktam Sajedi. To the thousands of undergraduates I have taught over the past two decades, thank you for teaching me how to teach, and for getting down in the soil pit and wading through the forest to look, touch, and listen to its wonders. I hope I have passed on to you some of my enthusiasm for what has always been so fascinating to me.

The colleagues I have had the pleasure of working with over the years are too numerous to list, but in particular I thank Drs. Dan Durall, Melanie Jones, and Randy Molina for sharing their enthusiasm for the underground life of forests. I thank Deborah DeLong for sharing a varied career in government and academia with me, our paths converging at the most intriguing times. I am grateful as well to the early days with my Forest Service colleagues in silviculture, especially Dave Coates and Teresa Newsome, and my early coauthor Jean Heineman.

I am grateful to my mentors and teachers for deepening my interest in the science of forests. My earliest advisor was Les Lavkulich, a pioneering soil chemist who showed me what it meant to be an outstanding teacher, who made soil genesis the most fascinating topic in the world, and who guided me through my bachelor's thesis. When I gained a position as a research silviculturist with the Forest Service in 1990, Alan Vyse took me under his wing and inspired me to learn scientific skills without losing sight of what makes a forest whole, and he gave me every opportunity to pursue graduate work in forest ecology. Alan, I am forever grateful for all that you taught me and the opportunities you provided. I am indebted to my master of science supervisor, Steve Radosevich, who brought the precise study of species interactions from the agricultural field to the forest, and who later saw that people were as important in plant communities as the plants themselves. I owe a heap of gratitude to my PhD supervisor, David A. Perry, who showed me how to understand forestry through the lens of ecology. I am proud to have been a student of all of you.

I am grateful to my collaborations with the many artists, writers, and filmmakers who have taken an interest in my work and cast it in a light for more to see. In particular, I am grateful to Lorraine Roy for creating the *Woven Woods,* Louie Schwartzberg for *Fantastic Fungi,* Richard Powers for *The Overstory,* Erna Buffie for *Smarty Plants,* and Dan McKinney and Julia Dordel for *Mother Trees Connect the Forest.* It has been a great pleasure to collaborate with my brother-in-law Bill Heath in bringing my work to the TED stage and making documentary films of the Mother Tree and Salmon Forest Projects, and for creating the archive of historic photographs of my family and life, some of which made it to these pages.

None of the work in this book would have happened without the funding and support of several institutions, granting agencies, and foundations. These include the British Columbia Ministry of Forests and Range, the University of British Columbia, the Natural Sciences and Engineering Research Council of Canada (NSERC), the Canadian Foundation for Innovation (CFI), Genome BC, the Forest Enhancement Society of British Columbia (FESBC), the Forest Carbon Initiative (FCI), and others. I also greatly appreciate the generous support of the Donner Canadian Foundation for the Salmon Forest Project and the Jena and Michael King Foundation for the Mother Tree Project.

Several important people read and commented on the manuscript and provided exceedingly helpful feedback. These included June Simard, Peter Simard, Robyn Simard, Bill Heath, Don Sachs, Trish Miller, Jean Roach, and Alan Vyse. I am also very grateful to Dr. Teresa "Sm'hayetsk" Ryan (Tsimshian Nation) for reviewing the content about the Aboriginal people, for teaching me about the Aboriginal way of seeing the world, and for seeing the value of connecting these small scientific findings with the deeper socio-ecological linkages that are fundamental to the indigenous ways of life. I am grateful to Nora Reichard, production editor at Penguin Random House, for her careful production editing of the manuscript.

I'm thankful for collaborations and discussions with members of the Coast Salish, Heiltsuk, Tsimshian, Haida, Athabascan, Interior Salish, and Ktunaxa First Peoples, on whose traditional, ancestral, and unceded territories we have lived and conducted this work.

I thank Don for being with me during some of the most difficult times, and the most joyous times too, and for being a wonderful father to our beautiful daughters, Hannah Rebekah Sachs and Nava Sophia Sachs. I have always been grateful for your love and support.

Finally, thank you, Mary, for always picking up my pieces, and for warily being ready for the next adventure.

I am responsible for the final content of this book. I have tried to be an honest broker of history, but sometimes I had to creatively fill in the gaps of my memory, or make small changes to protect the privacy of individuals. Some names have been left out for brevity, or changed to protect privacy, but I hope I have given credit where credit is due. For my students and colleagues, even where I did not mention your name, or used only your first name, I have cited your crucial work in Critical Sources.

CRITICAL SOURCES

INTRODUCTION: CONNECTIONS

Enderby and District Museum and Archives Historical Photograph Collection. *Log chute at falls near Mabel Lake in Winter. 1898.* (Located near Simard Creek on the east shore of Mabel Lake.) www.enderbymuseum.ca/archives.php.

Pierce, Daniel. 2018. 25 years after the war in the woods: Why B.C.'s forests are still in crisis. *The Narwhal.* https://thenarwhal.ca/25-years-after-clayoquot-sound-blockades-the-war-in-the-woods-never-ended-and-its-heating-back-up/.

Raygorodetsky, Greg. 2014. Ancient woods. Chapter 3 in *Everything Is Connected.* National Geographic. https://blog.nationalgeographic.org/2014/04/22/everything-is-connected-chapter-3-ancient-woods/.

Simard, Isobel. 1977. The Simard story. In *Flowing Through Time: Stories of Kingfisher and Mabel Lake.* Kingfisher History Committee, 321–22.

UBC Faculty of Forestry Alumni Relations and Development. Welcome forestry alumni. https://getinvolved.forestry.ubc.ca/alumni/.

Western Canada Wilderness Committee. 1985. Massive clearcut logging is ruining Clayoquot Sound. *Meares Island,* 2–3.

1: GHOSTS IN THE FOREST

Ashton, M. S., and Kelty, M. J. 2019. *The Practice of Silviculture: Applied Forest Ecology,* 10th ed. Hoboken, NJ: Wiley.

Edgewood Inonoaklin Women's Institute. 1991. *Just Where Is Edgewood?* Edgewood, BC: Edgewood History Book Committee, 138–41.

Hosie, R. C. 1979. *Native Trees of Canada,* 8th ed. Markham, ON: Fitzhenry & Whiteside Ltd.

Kimmins, J. P. 1996. *Forest Ecology: A Foundation for Sustainable Management,* 3rd ed. Upper Saddle River, NJ: Pearson Education.

Klinka, K., Worrall, J., Skoda, L., and Varga, P. 1999. *The Distribution and Synopsis of Ecological and Silvical Characteristics of Tree Species in British Columbia's Forests,* 2nd ed. Coquitlam, BC: Canadian Cartographics Ltd.

Ministry of Forest Act. 1979. *Revised Statutes of British Columbia.* Victoria, BC: Queen's Printer.

Ministry of Forests. 1980. *Forest and Range Resource Analysis Technical Report.* Victoria, BC: Queen's Printer.

National Audubon Society. 1981. *Field Guide to North American Mushrooms.* New York: Knopf.

Pearkes, Eileen Delehanty. 2016. *A River Captured: The Columbia River Treaty and Catastrophic Challenge.* Calgary, AB: Rocky Mountain Books.

Pojar, J., and MacKinnon, A. 2004. *Plants of Coastal British Columbia,* rev. ed. Vancouver, BC: Lone Pine Publishing.

Stamets, Paul. 2005. *Mycelium Running: How Mushrooms Can Save the World.* Berkeley, CA: Ten Speed Press.

Vaillant, John. 2006. *The Golden Spruce: A True Story of Myth, Madness and Greed.* Toronto: Vintage Canada.

Weil, R. R., and Brady, N. C. 2016. *The Nature and Properties of Soils,* 15th ed. Upper Saddle River, NJ: Pearson Education.

2: HAND FALLERS

Enderby and District Museum and Archives Historical Photograph Collection. *Henry Simard, Wilfred Simard, and a third unknown man breaking up a log jam in the Skookumchuck Rapids on part of a log drive down the Shuswap River. 1925.* www.enderbymuseum.ca/archives.php.

———. *Moving Simard's houseboat on Mabel Lake. 1925.* www.enderbymuseum.ca/archives.php.

Hatt, Diane. 1989. Wilfred and Isobel Simard. In *Flowing Through Time: Stories of Kingfisher and Mabel Lake.* Kingfisher History Committee, 323–24.

Mitchell, Hugh. 2014. Memories of Henry Simard. In *Flowing Through Time: Stories of Kingfisher and Mabel Lake.* Kingfisher History Committee, 325.

Oliver, C. D., and Larson, B. C. 1996. *Forest Stand Dynamics,* updated ed. New York: Wiley.

Pearase, Jackie. 2014. Jack Simard: A life in the Kingfisher. In *Flowing Through Time: Stories of Kingfisher and Mabel Lake.* Kingfisher History Committee, 326–28.

Soil Classification Working Group. 1998. *The Canadian System of Soil Classification,* 3rd ed. Agriculture and Agri-Food Canada Publication 1646. Ottawa, ON: NRC Research Press.

3: PARCHED

Arora, David. 1986. *Mushrooms Demystified,* 2nd ed. Berkeley, CA: Ten Speed Press.

British Columbia Ministry of Forests. 1991. *Ecosystems of British Columbia.* Special Report Series 6. Victoria, BC: BC Ministry of Forests. http://www.for.gov.bc.ca/hfd/pubs/Docs/Srs/SRseries.htm.

Burns, R. M., and Honkala, B. H., coord. 1990. *Silvics of North America.* Vol. 1, *Conifers.* Vol. 2, *Hardwoods.* USDA Agriculture Handbook 654. Washington, DC: U.S. Forest Service. Only available online at http://www.na.fs.fed.us/spfo/pubs/silvics%5Fmanual.

Parish, R., Coupe, R., and Lloyd, D. 1999. *Plants of Southern Interior British Columbia,* 2nd ed. Vancouver, BC: Lone Pine Publishing.

Pati, A. J. 2014. *Formica integroides* of Swakum Mountain: A qualitative and quantitative assessment and narrative of *Formica* mounding behaviors influencing litter decomposition in a dry, interior Douglas-fir forest in British Columbia. Master of science thesis, University of British Columbia. DOI: 10.14288/1.0166984.

4: TREED

Bjorkman, E. 1960. *Monotropa hypopitys* L.—An epiparasite on tree roots. *Physiologia Plantarum* 13: 308–27.

Fraser Basin Council. 2013. *Bridge Between Nations.* Vancouver, BC: Fraser Basin Council and Simon Fraser University.

Herrero, S. 2018. *Bear Attacks: Their Causes and Avoidance,* 3rd ed. Lanham, MD: Lyons Press.

Martin, K., and Eadie, J. M. 1999. Nest webs: A community wide approach to the management and conservation of cavity nesting birds. *Forest Ecology and Management* 115: 243–57.

M'Gonigle, Michael, and Wickwire, Wendy. 1988. *Stein: The Way of the River.* Vancouver, BC: Talonbooks.

Perry, D. A., Oren, R., and Hart, S. C. 2008. *Forest Ecosystems,* 2nd ed. Baltimore: The Johns Hopkins University Press.

Prince, N. 2002. Plateau fishing technology and activity: Stl'atl'imx, Secwepemc and Nlaka'pamux knowledge. In *Putting Fishers' Knowledge to Work,* ed. N. Haggan, C. Brignall, and L. J. Wood. Conference proceedings, August 27–30, 2001. *Fisheries Centre Research Reports* 11 (1): 381–91.

Smith, S., and Read, D. 2008. *Mycorrhizal Symbiosis*. London: Academic Press.

Swinomish Indian Tribal Community. 2010. *Swinomish Climate Change Initiative: Climate Adaptation Action Plan.* La Conner, WA: Swinomish Indian Tribal Community. http://www.swinomish-nsn.gov/climate_change/climate_main.html.

Thompson, D., and Freeman, R. 1979. *Exploring the Stein River Valley.* Vancouver, BC: Douglas & McIntyre.

Walmsley, M., Utzig, G., Vold, T., et al. 1980. *Describing Ecosystems in the Field.* RAB Technical Paper 2; Land Management Report 7. Victoria, BC: Research Branch, British Columbia Ministry of Environment, and British Columbia Ministry of Forests.

Wickwire, W. C. 1991. Ethnography and archaeology as ideology: The case of the Stein River valley. *BC Studies* 91–92: 51–78.

Wilson, M. 2011. Co-management re-conceptualized: Human-land relations in the Stein Valley, British Columbia. BA thesis, University of Victoria.

York, A., Daly, R., and Arnett, C. 2019. *They Write Their Dreams on the Rock Forever: Rock Writings in the Stein River Valley of British Columbia,* 2nd ed. Vancouver, BC: Talonbooks.

5: KILLING SOIL

British Columbia Ministry of Forests. 1986. *Silviculture Manual.* Victoria, BC: Silviculture Branch.

————. 1987. *Forest Amendment Act (No. 2).* Victoria, BC: Queen's Printer. This act enabled enforcement of silvicultural performance and shifted cost and responsibility for reforestation to companies harvesting timber.

British Columbia Parks. 2000. *Management Plan for Stein Valley Nlaka'pamux Heritage Park.* Kamloops: British Columbia Ministry of Environment, Lands and Parks, Parks Division.

Chazan, M., Helps, L., Stanley, A., and Thakkar, S., eds. 2011. *Home and Native Land: Unsettling Multiculturalism in Canada.* Toronto, ON: Between the Lines.

Dunford, M. P. 2002. The Simpcw of the North Thompson. *British Columbia Historical News* 25 (3): 6–8.

First Nations land rights and environmentalism in British Columbia. http://www.first nations.de/indian_land.htm.

Haeussler, S., and Coates, D. 1986. *Autecological Characteristics of Selected Species That Compete with Conifers in British Columbia: A Literature Review.* BC Land Management Report 33. Victoria, BC: BC Ministry of Forests.

Ignace, Ron. 2008. Our oral histories are our iron posts: Secwepemc stories and historical consciousness. PhD thesis, Simon Fraser University.

Lindsay, Bethany. 2018. "It blows my mind": How B.C. destroys a key natural wildfire defence every year. CBC News, Nov. 17, 2018. https://www.cbc.ca/news/canada /british-columbia/it-blows-my-mind-how-b-c-destroys-a-key-natural-wildfire -defence-every-year-1.4907358.

Malik, N., and Vanden Born, W. H. 1986. *Use of Herbicides in Forest Management.* Information Report NOR-X-282. Edmonton: Canadian Forestry Service.

Mather, J. 1986. *Assessment of Silviculture Treatments Used in the IDF Zone in the Western Kamloops Forest Region.* Kamloops: BC Ministry of Forestry Research Section, Kamloops Forest Region.

Nelson, J. 2019. Monsanto's rain of death on Canada's forests. Global Research. https:// www.globalresearch.ca/monsantos-rain-death-forests/5677614.

Simard, S. W. 1996. Design of a birch/conifer mixture study in the southern interior of British Columbia. In *Designing Mixedwood Experiments: Workshop Proceedings, March 2, 1995, Richmond, BC,* ed. P. G. Comeau and K. D. Thomas. Working Paper 20. Victoria, BC: Research Branch, BC Ministry of Forests, 8–11.

———. 1996. Mixtures of paper birch and conifers: An ecological balancing act. In *Silviculture of Temperate and Boreal Broadleaf-Conifer Mixtures: Proceedings of a Workshop Held Feb. 28–March 1, 1995, Richmond, BC,* ed. P. G. Comeau and K. D. Thomas. BC Ministry of Forests Land Management Handbook 36. Victoria, BC: BC Ministry of Forests, 15–21.

———. 1997. Intensive management of young mixed forests: Effects on forest health. In *Proceedings of the 45th Western International Forest Disease Work Conference, Sept. 15–19, 1997,* ed. R. Sturrock. Prince George, BC: Pacific Forestry Centre, 48–54.

———. 2009. Response diversity of mycorrhizas in forest succession following disturbance. Chapter 13 in *Mycorrhizas: Functional Processes and Ecological Impacts,* ed. C. Azcon-Aguilar, J. M. Barea, S. Gianinazzi, and V. Gianinazzi-Pearson. Heidelberg: Springer-Verlag, 187–206.

Simard, S. W., and Heineman, J. L. 1996. *Nine-Year Response of Douglas-Fir and the Mixed Hardwood-Shrub Complex to Chemical and Manual Release Treatments on an ICHmw2 Site Near Salmon Arm.* FRDA Research Report 257. Victoria, BC: Canadian Forest Service and BC Ministry of Forests.

———. 1996. *Nine-Year Response of Engelmann Spruce and the Willow Complex to Chemical and Manual Release Treatments on an Ichmw2 Site Near Vernon.* FRDA Research Report 258. Victoria, BC: Canadian Forest Service and BC Ministry of Forests.

———. 1996. *Nine-Year Response of Lodgepole Pine and the Dry Alder Complex to Chemical and Manual Release Treatments on an Ichmk1 Site Near Kelowna.* FRDA Research Report 259. Victoria, BC: Canadian Forest Service and BC Ministry of Forests.

Simard, S. W., Heineman, J. L., and Youwe, P. 1998. *Effects of Chemical and Manual Brushing on Conifer Seedlings, Plant Communities and Range Forage in the Southern*

Interior of British Columbia: Nine-Year Response. Land Management Report 45. Victoria, BC: BC Ministry of Forests.

Swanson, F., and Franklin, J. 1992. New principles from ecosystem analysis of Pacific Northwest forests. *Ecological Applications* 2: 262–74.

Wang, J. R., Zhong, A. L., Simard, S. W., and Kimmins, J. P. 1996. Aboveground biomass and nutrient accumulation in an age sequence of paper birch (*Betula papyrifera*) stands in the Interior Cedar Hemlock zone, British Columbia. *Forest Ecology and Management* 83: 27–38.

6: ALDER SWALES

Arnebrant, K., Ek, H., Finlay, R. D., and Söderström, B. 1993. Nitrogen translocation between *Alnus glutinosa* (L.) Gaertn. seedlings inoculated with *Frankia* sp. and *Pinus contorta* Doug, ex Loud seedlings connected by a common ectomycorrhizal mycelium. *New Phytologist* 124: 231–42.

Bidartondo, M. I., Redecker, D., Hijri, I., et al. 2002. Epiparasitic plants specialized on arbuscular mycorrhizal fungi. *Nature* 419: 389–92.

British Columbia Ministry of Forests, Lands and Natural Resources Operations. 1911–2012. Annual Service Plant Reports/Annual Reports. Victoria, BC: Crown Publications, www.for.gov.bc.ca/mof/annualreports.htm.

Brooks, J. R., Meinzer, F. C., Warren, J. M., et al. 2006. Hydraulic redistribution in a Douglas-fir forest: Lessons from system manipulations. *Plant, Cell and Environment* 29: 138–50.

Carpenter, C. V., Robertson, L. R., Gordon, J. C., and Perry, D. A. 1982. The effect of four new *Frankia* isolates on growth and nitrogenase activity in clones of *Alnus rubra* and *Alnus sinuata*. *Canadian Journal of Forest Research* 14: 701–6.

Cole, E. C., and Newton, M. 1987. Fifth-year responses of Douglas fir to crowding and non-coniferous competition. *Canadian Journal of Forest Research* 17: 181–86.

Daniels, L. D., Yocom, L. L., Sherriff, R. L., and Heyerdahl, E. K. 2018. Deciphering the complexity of historical hire regimes: Diversity among forests of western North America. In *Dendroecology,* ed. M. M. Amoroso et al. Ecological Studies vol. 231. New York: Springer International Publishing AG. DOI 10.1007/978-3-319-61669-8_8.

Hessburg, P. F., Miller, C. L., Parks, S. A., et al. 2019. Climate, environment, and disturbance history govern resilience of western North American forests. *Frontiers in Ecology and Evolution* 7: 239.

Ingham, R. E., Trofymow, J. A., Ingham, E. R., and Coleman, D. C. 1985. Interactions of bacteria, fungi, and their nematode grazers: Effects on nutrient cycling and plant growth. *Ecological Monographs* 55: 119–40.

Klironomos, J. N., and Hart, M. M. 2001. Animal nitrogen swap for plant carbon. *Nature* 410: 651–52.

Querejeta, J., Egerton-Warburton, L. M., and Allen, M. F. 2003. Direct nocturnal water transfer from oaks to their mycorrhizal symbionts during severe soil drying. *Oecologia* 134: 55–64.

Radosevich, S. R., and Roush, M. L. 1990. The role of competition in agriculture. In *Perspectives on Plant Competition,* ed. J. B. Grace and D. Tilman. San Diego, CA: Academic Press, Inc.

Sachs, D. L. 1991. *Calibration and initial testing of FORECAST for stands of lodgepole pine and Sitka alder in the interior of British Columbia.* Report 035-510-07403. Victoria, BC: British Columbia Ministry of Forests.

Simard, S. W. 1989. Competition among lodgepole pine seedlings and plant species in a Sitka alder dominated shrub community in the southern interior of British Columbia. Master of science thesis, Oregon State University.

———. 1990. *Competition between Sitka alder and lodgepole pine in the Montane Spruce zone in the southern interior of British Columbia.* FRDA Report 150. Victoria: BC: Forestry Canada and BC Ministry of Forests, 150.

Simard, S. W., Radosevich, S. R., Sachs, D. L., and Hagerman, S. M. 2006. Evidence for competition/facilitation trade-offs: Effects of Sitka alder density on pine regeneration and soil productivity. *Canadian Journal of Forest Research* 36: 1286–98.

Simard, S. W., Roach, W. J., Daniels, L. D., et al. Removal of neighboring vegetation predisposes planted lodgepole pine to growth loss during climatic drought and mortality from a mountain pine beetle infestation. In preparation.

Southworth, D., He, X. H., Swenson, W., et al. 2005. Application of network theory to potential mycorrhizal networks. *Mycorrhiza* 15: 589–95.

Wagner, R. G., Little, K. M., Richardson, B., and McNabb, K. 2006. The role of vegetation management for enhancing productivity of the world's forests. *Forestry* 79 (1): 57–79.

Wagner, R. G., Peterson, T. D., Ross, D. W., and Radosevich, S. R. 1989. Competition thresholds for the survival and growth of ponderosa pine seedlings associated with woody and herbaceous vegetation. *New Forests* 3: 151–70.

Walstad, J. D., and Kuch, P. J., eds. 1987. *Forest Vegetation Management for Conifer Production.* New York: John Wiley and Sons, Inc.

7: BAR FIGHT

Frey, B., and Schüepp, H. 1992. Transfer of symbiotically fixed nitrogen from berseem (*Trifolium alexandrinum* L.) to maize via vesicular-arbuscular mycorrhizal hyphae. *New Phytologist* 122: 447–54.

Haeussler, S., Coates, D., and Mather, J. 1990. *Autecology of common plants in British Columbia: A literature review.* FRDA Report 158. Victoria, BC: Forestry Canada and BC Ministry of Forests.

Heineman, J. L., Sachs, D. L., Simard, S. W., and Mather, W. J. 2010. Climate and site characteristics affect juvenile trembling aspen development in conifer plantations across southern British Columbia. *Forest Ecology & Management* 260: 1975–84.

Heineman, J. L., Simard, S. W., Sachs, D. L., and Mather, W. J. 2005. Chemical, grazing, and manual cutting treatments in mixed herb-shrub communities have no effect on interior spruce survival or growth in southern interior British Columbia. *Forest Ecology and Management* 205: 359–74.

———. 2007. Ten-year responses of Engelmann spruce and a high elevation Ericaceous shrub community to manual cutting treatments in southern interior British Columbia. *Forest Ecology and Management* 248: 153–62.

———. 2009. Trembling aspen removal effects on lodgepole pine in southern interior British Columbia: 10-year results. *Western Journal of Applied Forestry* 24: 17–23.

Miller, S. L., Durall, D. M., and Rygiewicz, P. T. 1989. Temporal allocation of ^{14}C to extramatrical hyphae of ectomycorrhizal ponderosa pine seedlings. *Tree Physiology* 5: 239–49.

Molina, R., Massicotte, H., and Trappe, J. M. 1992. Specificity phenomena in mycorrhizal symbiosis: Community-ecological consequences and practical implications.

In *Mycorrhizal Functioning: An Integrative Plant-Fungal Process*, ed. M. F. Allen. New York: Chapman and Hall, 357–423.

Morrison, D., Merler, H., and Norris, D. 1991. *Detection, recognition and management of Armillaria and Phellinus root diseases in the southern interior of British Columbia*. FRDA Report 179. Victoria, BC: Forestry Canada and BC Ministry of Forests.

Perry, D. A., Margolis, H., Choquette, C., et al. 1989. Ectomycorrhizal mediation of competition between coniferous tree species. *New Phytologist* 112: 501–11.

Rolando, C. A., Baillie, B. R., Thompson, D. G., and Little, K. M. 2007. The risks associated with glyphosate-based herbicide use in planted forests. *Forests* 8: 208.

Sachs, D. L., Sollins, P., and Cohen, W. B. 1998. Detecting landscape changes in the interior of British Columbia from 1975 to 1992 using satellite imagery. *Canadian Journal of Forest Research* 28: 23–36.

Simard, S. W. 1993. *PROBE: Protocol for operational brushing evaluations (first approximation)*. Land Management Report 86. Victoria, BC: BC Ministry of Forests.

———. 1995. *PROBE: Vegetation management monitoring in the southern interior of B.C.* Northern Interior Vegetation Management Association, Annual General Meeting, Jan. 18, 1995, Williams Lake, BC.

Simard, S. W., Heineman, J. L., Hagerman, S. M., et al. 2004. Manual cutting of Sitka alder-dominated plant communities: Effects on conifer growth and plant community structure. *Western Journal of Applied Forestry* 19: 277–87.

Simard, S. W., Heineman, J. L., Mather, W. J., et al. 2001. *Brushing effects on conifers and plant communities in the southern interior of British Columbia: Summary of PROBE results 1991–2000*. Extension Note 58. Victoria, BC: BC Ministry of Forestry.

Simard, S. W., Jones, M. D., Durall, D. M., et al. 2003. Chemical and mechanical site preparation: Effects on *Pinus contorta* growth, physiology, and microsite quality on steep forest sites in British Columbia. *Canadian Journal of Forest Research* 33: 1495–515.

Thompson, D. G., and Pitt, D. G. 2003. A review of Canadian forest vegetation management research and practice. *Annals of Forest Science* 60: 559–72.

8: RADIOACTIVE

Brownlee, C., Duddridge, J. A., Malibari, A., and Read, D. J. 1983. The structure and function of mycelial systems of ectomycorrhizal roots with special reference to their role in forming inter-plant connections and providing pathways for assimilate and water transport. *Plant Soil* 71: 433–43.

Callaway, R. M. 1995. Positive interactions among plants. *Botanical Review* 61 (4): 306–49.

Finlay, R. D., and Read, D. J. 1986. The structure and function of the vegetative mycelium of ectomycorrhizal plants. I. Translocation of ^{14}C-labelled carbon between plants interconnected by a common mycelium. *New Phytologist* 103: 143–56.

Francis, R., and Read, D. J. 1984. Direct transfer of carbon between plants connected by vesicular-arbuscular mycorrhizal mycelium. *Nature* 307: 53–56.

Jones, M. D., Durall, D. M., Harniman, S. M. K., et al. 1997. Ectomycorrhizal diversity on *Betula papyrifera* and *Pseudotsuga menziesii* seedlings grown in the greenhouse or outplanted in single-species and mixed plots in southern British Columbia. *Canadian Journal of Forest Research* 27: 1872–89.

McPherson, S. S. 2009. *Tim Berners-Lee: Inventor of the World Wide Web*. Minneapolis: Twenty-First Century Books.

Read, D. J., Francis, R., and Finlay, R. D. 1985. Mycorrhizal mycelia and nutrient cycling in plant communities. In *Ecological Interactions in Soil,* ed. A. H. Fitter, D. Atkinson, D. J. Read, and M. B. Usher. Oxford: Blackwell Scientific, 193–217.

Ryan, M. G., and Asao, S. 2014. Phloem transport in trees. *Tree Physiology* 34: 1–4.

Simard, S. W. 1990. *A retrospective study of competition between paper birch and planted Douglas-fir.* FRDA Report 147. Victoria, BC: Forestry Canada and BC Ministry of Forests.

Simard, S. W., Molina, R., Smith, J. E., et al. 1997. Shared compatibility of ectomycorrhizae on *Pseudotsuga menziesii* and *Betula papyrifera* seedlings grown in mixture in soils from southern British Columbia. *Canadian Journal of Forest Research* 27: 331–42.

Simard, S. W., Perry, D. A., Jones, M. D., et al. 1997. Net transfer of carbon between tree species with shared ectomycorrhizal fungi. *Nature* 388: 579–82.

Simard, S. W., and Vyse, A. 1992. *Ecology and management of paper birch and black cottonwood.* Land Management Report 75. Victoria, BC: BC Ministry of Forests.

9: QUID PRO QUO

Baleshta, K. E. 1998. The effect of ectomycorrhizae hyphal links on interactions between *Pseudotsuga menziesii* (Mirb.) Franco and *Betula papyrifera* Marsh. seedlings. Bachelors of natural resource sciences thesis, University College of the Cariboo.

Baleshta, K. E., Simard, S. W., Guy, R. D., and Chanway, C. P. 2005. Reducing paper birch density increases Douglas-fir growth and Armillaria root disease incidence in southern interior British Columbia. *Forest Ecology and Management* 208: 1–13.

Baleshta, K. E., Simard, S. W., and Roach, W. J. 2015. Effects of thinning paper birch on conifer productivity and understory plant diversity. *Scandinavian Journal of Forest Research* 30: 699–709.

DeLong, R., Lewis, K. J., Simard, S. W., and Gibson, S. 2002. Fluorescent pseudomonad population sizes baited from soils under pure birch, pure Douglas-fir and mixed forest stands and their antagonism toward *Armillaria ostoyae* in vitro. *Canadian Journal of Forest Research* 32: 2146–59.

Durall, D. M., Gamiet, S., Simard, S. W., et al. 2006. Effects of clearcut logging and tree species composition on the diversity and community composition of epigeous fruit bodies formed by ectomycorrhizal fungi. *Canadian Journal of Botany* 84: 966–80.

Fitter, A. H., Graves, J. D., Watkins, N. K., et al. 1998. Carbon transfer between plants and its control in networks of arbuscular mycorrhizas. *Functional Ecology* 12: 406–12.

Fitter, A. H., Hodge, A., Daniell, T. J., and Robinson, D. 1999. Resource sharing in plant-fungus communities: Did the carbon move for you? *Trends in Ecology and Evolution* 14: 70–71.

Kimmerer, Robin Wall. 2015. *Braiding Sweetgrass: Indigenous Wisdom, Scientific Knowledge and the Teachings of Plants.* Minneapolis: Milkweed Editions.

Perry, D. A. 1998. A moveable feast: The evolution of resource sharing in plant-fungus communities. *Trends in Ecology and Evolution* 13: 432–34.

———. 1999. Reply from D. A. Perry. *Trends in Ecology and Evolution* 14: 70–71.

Philip, Leanne. 2006. The role of ectomycorrhizal fungi in carbon transfer within common mycorrhizal networks. PhD dissertation, University of British Columbia. https://open.library.ubc.ca/collections/ubctheses/831/items/1.0075066.

Sachs, D. L. 1996. Simulation of the growth of mixed stands of Douglas-fir and paper birch using the FORECAST model. In *Silviculture of Temperate and Boreal Broadleaf-Conifer Mixtures: Proceedings of a Workshop Held Feb. 28–March 1, 1995,*

Richmond, BC, ed. P. G. Comeau and K. D. Thomas. BC Ministry of Forests Land Management Handbook 36. Victoria, BC: BC Ministry of Forests, 152–58.

Simard, S. W., and Durall, D. M. 2004. Mycorrhizal networks: A review of their extent, function and importance. *Canadian Journal of Botany* 82: 1140–65.

Simard, S. W., Durall, D. M., and Jones, M. D. 1997. Carbon allocation and carbon transfer between *Betula papyrifera* and *Pseudotsuga menziesii* seedlings using a ^{13}C pulse-labeling method. *Plant and Soil* 191: 41–55.

Simard, S. W., and Hannam, K. D. 2000. Effects of thinning overstory paper birch on survival and growth of interior spruce in British Columbia: Implications for reforestation policy and biodiversity. *Forest Ecology and Management* 129: 237–51.

Simard, S. W., Jones, M. D., and Durall, D. M. 2002. Carbon and nutrient fluxes within and between mycorrhizal plants. In *Mycorrhizal Ecology,* ed. M. van der Heijden and I. Sanders. Heidelberg: Springer-Verlag, 33–61.

Simard, S. W., Jones, M. D., Durall, D. M., et al. 1997. Reciprocal transfer of carbon isotopes between ectomycorrhizal *Betula papyrifera* and *Pseudotsuga menziesii. New Phytologist* 137: 529–42.

Simard, S. W., Perry, D. A., Smith, J. E., and Molina, R. 1997. Effects of soil trenching on occurrence of ectomycorrhizae on *Pseudotsuga menziesii* seedlings grown in mature forests of *Betula papyrifera* and *Pseudotsuga menziesii. New Phytologist* 136: 327–40.

Simard, S. W., and Sachs, D. L. 2004. Assessment of interspecific competition using relative height and distance indices in an age sequence of seral interior cedar-hemlock forests in British Columbia. *Canadian Journal of Forest Research* 34: 1228–40.

Simard, S. W., Sachs, D. L., Vyse, A., and Blevins, L. L. 2004. Paper birch competitive effects vary with conifer tree species and stand age in interior British Columbia forests: Implications for reforestation policy and practice. *Forest Ecology and Management* 198: 55–74.

Simard, S. W., and Zimonick, B. J. 2005. Neighborhood size effects on mortality, growth and crown morphology of paper birch. *Forest Ecology and Management* 214: 251–69.

Twieg, B. D., Durall, D. M., and Simard, S. W. 2007. Ectomycorrhizal fungal succession in mixed temperate forests. *New Phytologist* 176: 437–47.

Wilkinson, D. A. 1998. The evolutionary ecology of mycorrhizal networks. *Oikos* 82: 407–10.

Zimonick, B. J., Roach, W. J., and Simard, S. W. 2017. Selective removal of paper birch increases growth of juvenile Douglas fir while minimizing impacts on the plant community. *Scandinavian Journal of Forest Research* 32: 708–16.

10: PAINTING ROCKS

Aukema, B. H., Carroll, A. L., Zhu, J., et al. 2006. Landscape level analysis of mountain pine beetle in British Columbia, Canada: Spatiotemporal development and spatial synchrony within the present outbreak. *Ecography* 29: 427–41.

Beschta, R. L., and Ripple, W. L. 2014. Wolves, elk, and aspen in the winter range of Jasper National Park, Canada. *Canadian Journal of Forest Research* 37: 1873–85.

Chavardes, R. D., Daniels, L. D., Gedalof, Z., and Andison, D. W. 2018. Human influences superseded climate to disrupt the 20th century fire regime in Jasper National Park, Canada. *Dendrochronologia* 48: 10–19.

Cooke, B. J., and Carroll, A. L. 2017. Predicting the risk of mountain pine beetle spread to eastern pine forests: Considering uncertainty in uncertain times. *Forest Ecology and Management* 396: 11–25.

Cripps, C. L., Alger, G., and Sissons, R. 2018. Designer niches promote seedling survival in forest restoration: A 7-year study of whitebark pine (*Pinus albicaulis*) seedlings in Waterton Lakes National Park. *Forests* 9 (8): 477.

Cripps, C., and Miller Jr., O. K. 1993. Ectomycorrhizal fungi associated with aspen on three sites in the north-central Rocky Mountains. *Canadian Journal of Botany* 71: 1414–20.

Fraser, E. C., Lieffers, V. J., and Landhäusser, S. M. 2005. Age, stand density, and tree size as factors in root and basal grafting of lodgepole pine. *Canadian Journal of Botany* 83: 983–88.

———. 2006. Carbohydrate transfer through root grafts to support shaded trees. *Tree Physiology* 26: 1019–23.

Gorzelak, M., Pickles, B. J., Asay, A. K., and Simard, S. W. 2015. Inter-plant communication through mycorrhizal networks mediates complex adaptive behaviour in plant communities. *Annals of Botany Plants* 7: plv050.

Hutchins, H. E., and Lanner, R. M. 1982. The central role of Clark's nutcracker in the dispersal and establishment of whitebark pine. *Oecologia* 55: 192–201.

Mattson, D. J., Blanchard, D. M., and Knight, R. R. 1991. Food habits of Yellowstone grizzly bears, 1977–1987. *Canadian Journal of Zoology* 69: 1619–29.

McIntire, E. J. B., and Fajardo, A. 2011. Facilitation within species: A possible origin of group-selected superorganisms. *American Naturalist* 178: 88–97.

Miller, R., Tausch, R., and Waicher, W. 1999. Old-growth juniper and pinyon woodlands. In *Proceedings: Ecology and Management of Pinyon-Juniper Communities Within the Interior West, September 15–18, 1997, Provo, UT*, comp. Stephen B. Monsen and Richard Stevens. Proc. RMRS-P-9. Ogden, UT: U.S. Department of Agriculture, Forest Service, Rocky Mountain Research Station.

Mitton, J. B., and Grant, M. C. 1996. Genetic variation and the natural history of quaking aspen. *BioScience* 46: 25–31.

Munro, Margaret. 1998. Weed trees are crucial to forest, research shows. *Vancouver Sun*, May 14, 1998.

Perkins, D. L. 1995. A dendrochronological assessment of whitebark pine in the Sawtooth Salmon River Region, Idaho. Master of science thesis, University of Arizona.

Perry, D. A. 1995. Self-organizing systems across scales. *Trends in Ecology and Evolution* 10: 241–44.

———. 1998. A moveable feast: The evolution of resource sharing in plant-fungus communities. *Trends in Ecology and Evolution* 13: 432–34.

Raffa, K. F., Aukema, B. H., Bentz, B. J., et al. 2008. Cross-scale drivers of natural disturbances prone to anthropogenic amplification: Dynamics of biome-wide bark beetle eruptions. *BioScience* 58: 501–17.

Ripple, W. J., Beschta, R. L., Fortin, J. K., and Robbins, C. T. 2014. Trophic cascades from wolves to grizzly bears in Yellowstone. *Journal of Animal Ecology* 83: 223–33.

Schulman, E. 1954. Longevity under adversity in conifers. *Science* 119: 396–99.

Seip, D. R. 1992. Factors limiting woodland caribou populations and their interrelationships with wolves and moose in southeastern British Columbia. *Canadian Journal of Zoology* 70: 1494–1503.

———. 1996. Ecosystem management and the conservation of caribou habitat in British Columbia. *Rangifer* special issue 10: 203–7.

Simard, S. W. 2009. Mycorrhizal networks and complex systems: Contributions of soil ecology science to managing climate change effects in forested ecosystems. *Canadian Journal of Soil Science* 89 (4): 369–82.

————. 2009. The foundational role of mycorrhizal networks in self-organization of interior Douglas-fir forests. *Forest Ecology and Management* 258S: S95–107.

Tomback, D. F. 1982. Dispersal of whitebark pine seeds by Clark's nutcracker: A mutualism hypothesis. *Journal of Animal Ecology* 51: 451–67.

Van Wagner, C. E., Finney, M. A., and Heathcott, M. 2006. Historical fire cycles in the Canadian Rocky Mountain parks. *Forest Science* 52: 704–17.

11: MISS BIRCH

Baldocchi, D. B., Black, A., Curtis, P. S., et al. 2005. Predicting the onset of net carbon uptake by deciduous forests with soil temperature and climate data: A synthesis of FLUXNET data. *International Journal of Biometeorology* 49: 377–87.

Bérubé, J. A., and Dessureault, M. 1988. Morphological characterization of *Armillaria ostoyae* and *Armillaria sinapina* sp. nov. *Canadian Journal of Botany* 66: 2027–34.

Bradley, R. L., and Fyles, J. W. 1995. Growth of paper birch (*Betula papyrifera*) seedlings increases soil available C and microbial acquisition of soil-nutrients. *Soil Biology and Biochemistry* 27: 1565–71.

British Columbia Ministry of Forests. 2000. *Establishment to Free Growing Guidebook*, rev. ed, version 2.2. Victoria, BC: British Columbia Ministry of Forests, Forest Practices Branch.

British Columbia Ministry of Forests and BC Ministry of Environment, Lands and Parks. 1995. *Root Disease Management Guidebook*. Victoria, BC: Forest Practices Code. http://www.for.gov.bc.ca/tasb/legsregs/fpc/fpcguide/root/roottoc.htm.

Castello, J. D., Leopold, D. J., and Smallidge, P. J. 1995. Pathogens, patterns, and processes in forest ecosystems. *BioScience* 45: 16–24.

Chanway, C. P., and Holl, F. B. 1991. Biomass increase and associative nitrogen fixation of mycorrhizal *Pinus contorta* seedlings inoculated with a plant growth promoting *Bacillus* strain. *Canadian Journal of Botany* 69: 507–11.

Cleary, M. R., Arhipova, N., Morrison, D. J., et al. 2013. Stump removal to control root disease in Canada and Scandinavia: A synthesis of results from long-term trials. *Forest Ecology and Management* 290: 5–14.

Cleary, M., van der Kamp, B., and Morrison, D. 2008. British Columbia's southern interior forests: Armillaria root disease stand establishment decision aid. *BC Journal of Ecosystems and Management* 9 (2): 60–65.

Coates, K. D., and Burton, P. J. 1999. Growth of planted tree seedlings in response to ambient light levels in northwestern interior cedar-hemlock forests of British Columbia. *Canadian Journal of Forest Research* 29: 1374–82.

Comeau, P. G., White, M., Kerr, G., and Hale, S. E. 2010. Maximum density-size relationships for Sitka spruce and coastal Douglas fir in Britain and Canada. *Forestry* 83: 461–68.

DeLong, D. L., Simard, S. W., Comeau, P. G., et al. 2005. Survival and growth responses of planted seedlings in root disease infected partial cuts in the Interior Cedar Hemlock zone of southeastern British Columbia. *Forest Ecology and Management* 206: 365–79.

Dixon, R. K., Brown, S., Houghton, R. A., et al. 1994. Carbon pools and flux of global forest ecosystems. *Science* 263: 185–91.

Fall, A., Shore, T. L., Safranyik, L., et al. 2003. Integrating landscape-scale mountain pine beetle projection and spatial harvesting models to assess management strategies. In *Mountain Pine Beetle Symposium: Challenges and Solutions. Oct. 30–31, 2003,*

Kelowna, British Columbia, ed. T. L. Shore, J. E. Brooks, and J. E. Stone. Information Report BC-X-399. Victoria, BC: Natural Resources Canada, Canadian Forest Service, Pacific Forestry Centre, 114–32.

Feurdean, A., Veski, S., Florescu, G., et al. 2017. Broadleaf deciduous forest counterbalanced the direct effect of climate on Holocene fire regime in hemiboreal/boreal region (NE Europe). *Quaternary Science Reviews* 169: 378–90.

Hély, C., Bergeron, Y., and Flannigan, M. D. 2000. Effects of stand composition on fire hazard in mixed-wood Canadian boreal forest. *Journal of Vegetation Science* 11: 813–24.

———. 2001. Role of vegetation and weather on fire behavior in the Canadian mixed-wood boreal forest using two fire behavior prediction systems. *Canadian Journal of Forest Research* 31: 430–41.

Hoekstra, J. M., Boucher, T. M., Ricketts, T. H., and Roberts, C. 2005. Confronting a biome crisis: Global disparities of habitat loss and protection. *Ecology Letters* 8: 23–29.

Hope, G. D. 2007. Changes in soil properties, tree growth, and nutrition over a period of 10 years after stump removal and scarification on moderately coarse soils in interior British Columbia. *Forest Ecology and Management* 242: 625–35.

Kinzig, A. P., Pacala, S., and Tilman, G. D., eds. 2002. *The Functional Consequences of Biodiversity: Empirical Progress and Theoretical Extensions.* Princeton: Princeton University Press.

Knohl, A., Schulze, E. D., Kolle, O., and Buchmann, N. 2003. Large carbon uptake by an unmanaged 250-year-old deciduous forest in Central Germany. *Agricultural and Forest Meteorology* 118: 151–67.

LePage, P., and Coates, K. D. 1994. Growth of planted lodgepole pine and hybrid spruce following chemical and manual vegetation control on a frost-prone site. *Canadian Journal of Forest Research* 24: 208–16.

Mann, M. E., Bradley, R. S., and Hughs, M. K. 1998. Global-scale temperature patterns and climate forcing over the past six centuries. *Nature* 392: 779–87.

Morrison, D. J., Wallis, G. W., and Weir, L. C. 1988. *Control of Armillaria and Phellinus root diseases: 20-year results from the Skimikin stump removal experiment.* Information Report BC x-302. Victoria, BC: Canadian Forest Service.

Newsome, T. A., Heineman, J. L., and Nemec, A. F. L. 2010. A comparison of lodgepole pine responses to varying levels of trembling aspen removal in two dry south-central British Columbia ecosystems. *Forest Ecology and Management* 259: 1170–80.

Simard, S. W., Beiler, K. J., Bingham, M. A., et al. 2012. Mycorrhizal networks: Mechanisms, ecology and modelling. *Fungal Biology Reviews* 26: 39–60.

Simard, S. W., Blenner-Hassett, T., and Cameron, I. R. 2004. Precommercial thinning effects on growth, yield and mortality in even-aged paper birch stands in British Columbia. *Forest Ecology and Management* 190: 163–78.

Simard, S. W., Hagerman, S. M., Sachs, D. L., et al. 2005. Conifer growth, *Armillaria ostoyae* root disease and plant diversity responses to broadleaf competition reduction in temperate mixed forests of southern interior British Columbia. *Canadian Journal of Forest Research* 35: 843–59.

Simard, S. W., Heineman, J. L., Mather, W. J., et al. 2001. *Effects of Operational Brushing on Conifers and Plant Communities in the Southern Interior of British Columbia: Results from PROBE 1991–2000.* BC Ministry of Forests and Land Management Handbook 48. Victoria, BC: BC Ministry of Forests.

Simard, S. W., and Vyse, A. 2006. Trade-offs between competition and facilitation: A case study of vegetation management in the interior cedar-hemlock forests of southern British Columbia. *Canadian Journal of Forest Research* 36: 2486–96.

van der Kamp, B. J. 1991. Pathogens as agents of diversity in forested landscapes. *Forestry Chronicle* 67: 353–54.

Vyse, A., Cleary, M. A., and Cameron, I. R. 2013. Tree species selection revisited for plantations in the Interior Cedar Hemlock zone of southern British Columbia. *Forestry Chronicle* 89: 382–91.

Vyse, A., and Simard, S. W. 2009. Broadleaves in the interior of British Columbia: Their extent, use, management and prospects for investment in genetic conservation and improvement. *Forestry Chronicle* 85: 528–37.

Weir, L. C., and Johnson, A. L. S. 1970. Control of *Poria weirii* study establishment and preliminary evaluations. Canadian Forest Service, Forest Research Laboratory, Victoria, Canada.

White, R. H., and Zipperer, W. C. 2010. Testing and classification of individual plants for fire behaviour: Plant selection for the wildland-urban interface. *International Journal of Wildland Fire* 19: 213–27.

12: NINE-HOUR COMMUTE

Babikova, Z., Gilbert, L., Bruce, T. J. A., et al. 2013. Underground signals carried through common mycelial networks warn neighbouring plants of aphid attack. *Ecology Letters* 16: 835–43.

Barker, J. S., Simard, S. W., and Jones, M. D. 2014. Clearcutting and wildfire have comparable effects on growth of directly seeded interior Douglas-fir. *Forest Ecology and Management* 331: 188–95.

Barker, J. S., Simard, S. W., Jones, M. D., and Durall, D. M. 2013. Ectomycorrhizal fungal community assembly on regenerating Douglas-fir after wildfire and clearcut harvesting. *Oecologia* 172: 1179–89.

Barto, E. K., Hilker, M., Müller, F., et al. 2011. The fungal fast lane: Common mycorrhizal networks extend bioactive zones of allelochemicals in soils. *PLOS ONE* 6: e27195.

Barto, E. K., Weidenhamer, J. D., Cipollini, D., and Rillig, M. C. 2012. Fungal superhighways: Do common mycorrhizal networks enhance below ground communication? *Trends in Plant Science* 17: 633–37.

Beiler, K. J., Durall, D. M., Simard, S.W., et al. 2010. Mapping the wood-wide web: Mycorrhizal networks link multiple Douglas-fir cohorts. *New Phytologist* 185: 543–53.

Beiler, K. J., Simard, S. W., and Durall, D. M. 2015. Topology of *Rhizopogon* spp. mycorrhizal meta-networks in xeric and mesic old-growth interior Douglas-fir forests. *Journal of Ecology* 103: 616–28.

Beiler, K. J., Simard, S. W., Lemay, V., and Durall, D. M. 2012. Vertical partitioning between sister species of *Rhizopogon* fungi on mesic and xeric sites in an interior Douglas-fir forest. *Molecular Ecology* 21: 6163–74.

Bingham, M. A., and Simard, S. W. 2011. Do mycorrhizal network benefits to survival and growth of interior Douglas-fir seedlings increase with soil moisture stress? *Ecology and Evolution* 3: 306–16.

———. 2012. Ectomycorrhizal networks of old *Pseudotsuga menziesii* var. *glauca* trees facilitate establishment of conspecific seedlings under drought. *Ecosystems* 15: 188–99.

———. 2012. Mycorrhizal networks affect ectomycorrhizal fungal community similarity between conspecific trees and seedlings. *Mycorrhiza* 22: 317–26.

———. 2013. Seedling genetics and life history outweigh mycorrhizal network potential to improve conifer regeneration under drought. *Forest Ecology and Management* 287: 132–39.

Carey, E. V., Marler, M. J., and Callaway, R. M. 2004. Mycorrhizae transfer carbon from a native grass to an invasive weed: Evidence from stable isotopes and physiology. *Plant Ecology* 172: 133–41.

Defrenne, C. A., Oka, G. A., Wilson, J. E., et al. 2016. Disturbance legacy on soil carbon stocks and stability within a coastal temperate forest of southwestern British Columbia. *Open Journal of Forestry* 6: 305–23.

Erland, L. A. E., Shukla, M. R., Singh, A. S., and Murch, S. J. 2018. Melatonin and serotonin: Mediators in the symphony of plant morphogenesis. *Journal of Pineal Research* 64: e12452.

Heineman, J. L., Simard, S. W., and Mather, W. J. 2002. *Natural regeneration of small patch cuts in a southern interior ICH forest.* Working Paper 64. Victoria, BC: BC Ministry of Forests.

Jones, M. D., Twieg, B., Ward, V., et al. 2010. Functional complementarity of Douglas-fir ectomycorrhizas for extracellular enzyme activity after wildfire or clearcut logging. *Functional Ecology* 4: 1139–51.

Kazantseva, O., Bingham, M. A., Simard, S. W., and Berch, S. M. 2009. Effects of growth medium, nutrients, water and aeration on mycorrhization and biomass allocation of greenhouse-grown interior Douglas-fir seedlings. *Mycorrhiza* 20: 51–66.

Kiers, E. T., Duhamel, M., Beesetty, Y., et al. 2011. Reciprocal rewards stabilize cooperation in the mycorrhizal symbiosis. *Science* 333: 880–82.

Kretzer, A. M., Dunham, S., Molina, R., and Spatafora, J. W. 2004. Microsatellite markers reveal the below ground distribution of genets in two species of *Rhizopogon* forming tuberculate ectomycorrhizas on Douglas fir. *New Phytologist* 161: 313–20.

Lewis, K., and Simard, S. W. 2012. Transforming forest management in B.C. Opinion editorial, special to the *Vancouver Sun,* March 11, 2012.

Marcoux, H. M., Daniels, L. D., Gergel, S. E., et al. 2015. Differentiating mixed- and high-severity fire regimes in mixed-conifer forests of the Canadian Cordillera. *Forest Ecology and Management* 341: 45–58.

Marler, M. J., Zabinski, C. A., and Callaway, R. M. 1999. Mycorrhizae indirectly enhance competitive effects of an invasive forb on a native bunchgrass. *Ecology* 80: 1180–86.

Mather, W. J., Simard, S. W., Heineman, J. L., Sachs, D. L. 2010. Decline of young lodgepole pine in southern interior British Columbia. *Forestry Chronicle* 86: 484–97.

Perry, D. A., Hessburg, P. F., Skinner, C. N., et al. 2011. The ecology of mixed severity fire regimes in Washington, Oregon, and Northern California, *Forest Ecology and Management* 262: 703–17.

Philip, L. J., Simard, S. W., and Jones, M. D. 2011. Pathways for belowground carbon transfer between paper birch and Douglas-fir seedlings. *Plant Ecology and Diversity* 3: 221–33.

Roach, W. J., Simard, S. W., and Sachs, D. L. 2015. Evidence against planting lodgepole pine monocultures in cedar-hemlock forests in southern British Columbia. *Forestry* 88: 345–58.

Schoonmaker, A. L., Teste, F. P., Simard, S. W., and Guy, R. D. 2007. Tree proximity, soil pathways and common mycorrhizal networks: Their influence on utilization of redistributed water by understory seedlings. *Oecologia* 154: 455–66.

Simard, S. W. 2009. The foundational role of mycorrhizal networks in self-organization of interior Douglas-fir forests. *Forest Ecology and Management* 258S: S95–107.

Simard, S. W., ed. 2010. *Climate Change and Variability*. Intech. https://www.intechopen.com/books/climate-change-and-variability.

———. 2012. Mycorrhizal networks and seedling establishment in Douglas-fir forests. Chapter 4 in *Biocomplexity of Plant-Fungal Interactions*, ed. D. Southworth. Ames, IA: Wiley-Blackwell, 85–107.

———. 2017. The mother tree. In *The Word for World Is Still Forest*, ed. Anna-Sophie Springer and Etienne Turpin. Berlin: K. Verlag and the Haus der Kulturen der Welt.

———. 2018. Mycorrhizal networks facilitate tree communication, learning and memory. Chapter 10 in *Memory and Learning in Plants,* ed. F. Baluska, M. Gagliano, and G. Witzany. West Sussex, UK: Springer, 191–213.

Simard, S. W., Asay, A. K., Beiler, K. J., et al. 2015. Resource transfer between plants through ectomycorrhizal networks. In *Mycorrhizal Networks*, ed. T. R. Horton. Ecological Studies vol. 224. Dordrecht: Springer, 133–76.

Simard, S. W., and Lewis, K. 2011. New policies needed to save our forests. Opinion editorial, special to the *Vancouver Sun,* April 8, 2011.

Simard, S. W., Martin, K., Vyse, A., and Larson, B. 2013. Meta-networks of fungi, fauna and flora as agents of complex adaptive systems. Chapter 7 in *Managing World Forests as Complex Adaptive Systems: Building Resilience to the Challenge of Global Change,* ed. K. Puettmann, C. Messier, and K. D. Coates. New York: Routledge, 133–64.

Simard, S. W., Mather, W. J., Heineman, J. L., and Sachs, D. L. 2010. Too much of a good thing? Planted lodgepole pine at risk of decline in British Columbia. *Silviculture Magazine* Winter 2010: 26–29.

Teste, F. P., Karst, J., Jones, M. D., et al. 2006. Methods to control ectomycorrhizal colonization: Effectiveness of chemical and physical barriers. *Mycorrhiza* 17: 51–65.

Teste, F. P., and Simard, S. W. 2008. Mycorrhizal networks and distance from mature trees alter patterns of competition and facilitation in dry Douglas-fir forests. *Oecologia* 158: 193–203.

Teste, F. P., Simard, S. W., and Durall, D. M. 2009. Role of mycorrhizal networks and tree proximity in ectomycorrhizal colonization of planted seedlings. *Fungal Ecology* 2: 21–30.

Teste, F. P., Simard, S. W., Durall, D. M., et al. 2010. Net carbon transfer occurs under soil disturbance between *Pseudotsuga menziesii* var. *glauca* seedlings in the field. *Journal of Ecology* 98: 429–39.

Teste, F. P., Simard, S. W., Durall, D. M., et al. 2009. Access to mycorrhizal networks and tree roots: Importance for seedling survival and resource transfer. *Ecology* 90: 2808–22.

Twieg, B., Durall, D. M., Simard, S. W., and Jones, M. D. 2009. Influence of soil nutrients on ectomycorrhizal communities in a chronosequence of mixed temperate forests. *Mycorrhiza* 19: 305–16.

Van Dorp, C. 2016. Rhizopogon mycorrhizal networks with interior Douglas fir in selectively harvested and non-harvested forests. Master of science thesis, University of British Columbia.

Vyse, A., Ferguson, C., Simard, S. W., et al. 2006. Growth of Douglas-fir, lodgepole pine, and ponderosa pine seedlings underplanted in a partially-cut, dry Douglas-fir stand in south-central British Columbia. *Forestry Chronicle* 82: 723–32.

Woods, A., and Bergerud, W. 2008. *Are free-growing stands meeting timber productivity expectations in the Lakes Timber supply area?* FREP Report 13. Victoria, BC: BC Ministry of Forests and Range, Forest Practices Branch.

Woods, A., Coates, K. D., and Hamann, A. 2005. Is an unprecedented *Dothistroma* needle blight epidemic related to climate change? *BioScience* 55 (9): 761–69.

Zabinski, C. A., Quinn, L., and Callaway, R. M. 2002. Phosphorus uptake, not carbon transfer, explains arbuscular mycorrhizal enhancement of *Centaurea maculosa* in the presence of native grassland species. *Functional Ecology* 16: 758–65.

Zustovic, M. 2012. The effects of forest gap size on Douglas-fir seedling establishment in the southern interior of British Columbia. Master of science thesis, University of British Columbia.

13: CORE SAMPLING

Aitken, S. N., Yeaman, S., Holliday, J. A., et al. 2008. Adaptation, migration or extirpation: Climate change outcomes for tree populations. *Evolutionary Applications* 1: 95–111.

D'Antonio, C. M., and Vitousek, P. M. 1992. Biological invasions by exotic grasses, the grass/fire cycle, and global change. *Annual Review of Ecology and Systematics* 23: 63–87.

Eason, W. R., and Newman, E. I. 1990. Rapid cycling of nitrogen and phosphorus from dying roots of *Lolium perenne*. *Oecologia* 82: 432.

Eason, W. R., Newman, E. I., and Chuba, P. N. 1991. Specificity of interplant cycling of phosphorus: The role of mycorrhizas. *Plant Soil* 137: 267–74.

Franklin, J. F., Shugart, H. H., and Harmon, M. E. 1987. Tree death as an ecological process: Causes, consequences and variability of tree mortality. *BioScience* 37: 550–56.

Hamann, A., and Wang, T. 2006. Potential effects of climate change on ecosystem and tree species distribution in British Columbia. *Ecology* 87: 2773–86.

Johnstone, J. F., Allen, C. D., Franklin, J. F., et al. 2016. Changing disturbance regimes, ecological memory, and forest resilience. *Frontiers in Ecology and the Environment* 14: 369–78.

Kesey, Ken. 1977. *Sometimes a Great Notion*. New York: Penguin Books.

Lotan, J. E., and Perry, D. A. 1983. *Ecology and Regeneration of Lodgepole Pine*. Agriculture Handbook 606. Missoula, MT: INTF&RES, USDA Forest Service.

Maclauchlan, L. E., Daniels, L. D., Hodge, J. C., and Brooks, J. E. 2018. Characterization of western spruce budworm outbreak regions in the British Columbia Interior. *Canadian Journal of Forest Research* 48: 783–802.

McKinney, D., and Dordel, J. 2011. *Mother Trees Connect the Forest* (video). http://www.karmatube.org/videos.php?id=2764.

Safranyik, L., and Carroll, A. L. 2006. The biology and epidemiology of the mountain pine beetle in lodgepole pine forests. Chapter 1 in *The Mountain Pine Beetle: A Synthesis of Biology, Management, and Impacts on Lodgepole Pine*, ed. L. Safranyik and W. R. Wilson. Victoria, BC: Natural Resources Canada, Canadian Forest Service, Pacific Forestry Centre, 3–66.

Song, Y. Y., Chen, D., Lu, K., et al. 2015. Enhanced tomato disease resistance primed by arbuscular mycorrhizal fungus. *Frontiers in Plant Science* 6: 1–13.

Song, Y. Y., Simard, S. W., Carroll, A., et al. 2015. Defoliation of interior Douglas-fir elicits carbon transfer and defense signalling to ponderosa pine neighbors through ectomycorrhizal networks. *Scientific Reports* 5: 8495.

Song, Y. Y., Ye, M., Li, C., et al. 2014. Hijacking common mycorrhizal networks for herbivore-induced defence signal transfer between tomato plants. *Scientific Reports* 4: 3915.

Song, Y. Y., Zeng, R. S., Xu, J. F., et al. 2010. Interplant communication of tomato plants through underground common mycorrhizal networks. *PLOS ONE* 5: e13324.

Taylor, S. W., and Carroll, A. L. 2004. Disturbance, forest age dynamics and mountain pine beetle outbreaks in BC: A historical perspective. In *Challenges and Solutions: Proceedings of the Mountain Pine Beetle Symposium. Kelowna, British Columbia, Canada, Oct. 30–31, 2003*, ed. T. L. Shore, J. E. Brooks, and J. E. Stone. Information Report BC-X-399. Victoria: Canadian Forest Service, Pacific Forestry Centre, 41–51.

14: BIRTHDAYS

Allen, C. D., Macalady, A. K., Chenchouni, H., et al. 2010. A global overview of drought and heat-induced tree mortality reveals emerging climate change risks for forests. *Forest Ecology and Management* 259: 660–84.

Asay, A. K. 2013. Mycorrhizal facilitation of kin recognition in interior Douglas-fir (*Pseudotsuga menziesii* var. *glauca*). Master of science thesis, University of British Columbia. DOI: 10.14288/1.0103374.

Bhatt, M., Khandelwal, A., and Dudley, S. A. 2011. Kin recognition, not competitive interactions, predicts root allocation in young *Cakile edentula* seedling pairs. *New Phytologist* 189: 1135–42.

Biedrzycki, M. L., Jilany, T. A., Dudley, S. A., and Bais, H. P. 2010. Root exudates mediate kin recognition in plants. *Communicative and Integrative Biology* 3: 28–35.

Brooker, R. W., Maestre, F. T., Callaway, R. M., et al. 2008. Facilitation in plant communities: The past, the present, and the future. *Journal of Ecology* 96: 18–34.

Donohue, K. 2003. The influence of neighbor relatedness on multilevel selection in the Great Lakes sea rocket. *American Naturalist* 162: 77–92.

Dudley, S. A., and File, A. L. 2007. Kin recognition in an annual plant. *Biology Letters* 3: 435–38.

File, A. L., Klironomos, J., Maherali, H., and Dudley, S. A. 2012. Plant kin recognition enhances abundance of symbiotic microbial partner. *PLOS ONE* 7: e45648.

Fontaine, S., Bardoux, G., Abbadie, L., and Mariotti, A. 2004. Carbon input to soil may decrease soil carbon content. *Ecology Letters* 7: 314–20.

Fontaine, S., Barot, S., Barré, P., et al. 2007. Stability of organic carbon in deep soil layers controlled by fresh carbon supply. *Nature* 450: 277–80.

Franklin, J. F., Cromack, K. Jr., Denison, W., et al. 1981. *Ecological characteristics of old-growth Douglas-fir forests*. General Technical Report PNW-GTR-118. Portland, OR: U.S. Department of Agriculture, Forest Service, Pacific Northwest Forest and Range Experiment Station.

Gilman, Dorothy. 1966. *The Unexpected Mrs. Pollifax*. New York: Fawcett.

Hamilton, W. D. 1964. The genetical evolution of social behaviour. *Journal of Theoretical Biology* 7: 1–16.

Harper, T. 2019. Breastless friends forever: How breast cancer brought four women together. *Nelson Star*, August 2, 2019. https://www.nelsonstar.com/community/breastless-friends-forever-how-breast-cancer-brought-four-women-together/.

Harte, J. 1996. How old is that old yew? *At the Edge* 4: 1–9.

Karban, R., Shiojiri, K., Ishizaki, S., et al. 2013. Kin recognition affects plant communication and defence. *Proceedings of the Royal Society B: Biological Sciences* 280: 20123062.

Luyssaert, S., Schulze, E. D., Börner, A., et al. 2008. Old-growth forests as global carbon sinks. *Nature* 455: 213–15.

Pickles, B. J., Twieg, B. D., O'Neill, G. A., et al. 2015. Local adaptation in migrated interior Douglas-fir seedlings is mediated by ectomycorrhizae and other soil factors. *New Phytologist* 207: 858–71.

Pickles, B. J., Wilhelm, R., Asay, A. K., et al. 2017. Transfer of ^{13}C between paired Douglas-fir seedlings reveals plant kinship effects and uptake of exudates by ectomycorrhizas. *New Phytologist* 214: 400–411.

Rehfeldt, G. E., Leites, L. P., St. Clair, J. B., et al. 2014. Comparative genetic responses to climate in the varieties of *Pinus ponderosa* and *Pseudotsuga menziesii*: Clines in growth potential. *Forest Ecology and Management* 324: 138–46.

Restaino, C. M., Peterson, D. L., and Littell, J. 2016. Increased water deficit decreases Douglas fir growth throughout western US forests. *Proceedings of the National Academy of Sciences* 113: 9557–62.

Simard, S. W. 2014. The networked beauty of forests. TED-Ed, New Orleans. https://ed.ted.com/lessons/the-networked-beauty-of-forests-suzanne-simard.

St. Clair, J. B., Mandel, N. L., and Vance-Borland, K. W. 2005. Genecology of Douglas fir in western Oregon and Washington. *Annals of Botany* 96: 1199–214.

Turner, N. J. 2008. *The Earth's Blanket: Traditional Teachings for Sustainable Living.* Seattle: University of Washington Press.

Turner, N. J., and Cocksedge, W. 2001. Aboriginal use of non-timber forest products in northwestern North America. *Journal of Sustainable Forestry* 13: 31–58.

Wall, M. E., and Wani, M. C. 1995. Camptothecin and taxol: Discovery to clinic— Thirteenth Bruce F. Cain Memorial Award Lecture. *Cancer Research* 55: 753–60.

15: PASSING THE WAND

Alila, Y., Kuras, P. K., Schnorbus, M., and Hudson, R. 2009. Forests and floods: A new paradigm sheds light on age-old controversies. American Geophysical Union. *Water Resources Research* 45: W08416.

Artelle, K. A., Stephenson, J., Bragg, C., et al. 2018. Values-led management: The guidance of place-based values in environmental relationships of the past, present, and future. *Ecology and Society* 23 (3): 35.

Asay, A. K. 2019. Influence of kin, density, soil inoculum potential and interspecific competition on interior Douglas-fir (*Pseudotsuga menziesii* var. *glauca*) performance and adaptive traits. PhD dissertation, University of British Columbia.

British Columbia Ministry of Forests and Range and British Columbia Ministry of Environment. 2010. *Field Manual for Describing Terrestrial Ecosystems,* 2nd ed. Land Management Handbook 25. Victoria, BC: Ministry of Forests and Range Research Branch.

Cox, Sarah. 2019. "You can't drink money": Kootenay communities fight logging to protect their drinking water. *The Narwhal.* https://thenarwhal.ca/you-cant-drink-money-kootenay-communities-fight-logging-protect-drinking-water/.

Gill, I. 2009. *All That We Say Is Ours: Guujaaw and the Reawakening of the Haida Nation.* Vancouver: Douglas & McIntyre.

Golder Associates. 2014. *Furry Creek detailed site investigations and human health and ecological risk assessment.* Vol. 1, *Methods and results.* Report 1014210038-501-R-RevO.

Gorzelak, M. A. 2017. Kin-selected signal transfer through mycorrhizal networks in Douglas-fir. PhD dissertation, University of British Columbia. DOI: 10.14288/1.0355225.

Harding, J. N., and Reynolds, J. D. 2014. Opposing forces: Evaluating multiple ecological roles of Pacific salmon in coastal stream ecosystems. *Ecosphere* 5: art157.

Hocking, M. D., and Reynolds, J. D. 2011. Impacts of salmon on riparian plant diversity. *Science* 331 (6024): 1609–12.

Kinzig, A. P., Ryan, P., Etienne, M., et al. 2006. Resilience and regime shifts: Assessing cascading effects. *Ecology and Society* 11: 20.

Kurz, W. A., Dymond, C. C., Stinson, G., et al. 2008. Mountain pine beetle and forest carbon: Feedback to climate change. *Nature* 452: 987–90.

Larocque, A. 2105. Forests, fish, fungi: Mycorrhizal associations in the salmon forests of BC. PhD proposal, University of British Columbia.

Louw, Deon. 2015. Interspecific interactions in mixed stands of paper birch (*Betula papyrifera*) and interior Douglas-fir (*Pseudotsuga mensiezii* var. *glauca*). Master of science thesis, University of British Columbia. https://open.library.ubc.ca/collections /ubctheses/24/items/1.0166375.

Marren, P., Marwan, H., and Alila, Y. 2013. Hydrological impacts of mountain pine beetle infestation: Potential for river channel changes. In *Cold and Mountain Region Hydrological Systems Under Climate Change: Towards Improved Projections, Proceedings of H02, IAHS-IAPSO-IASPEI Assembly, Gothenburg, Sweden, July 2013.* IAHS Publication 360: 77–82.

Mathews, D. L., and Turner, N. J. 2017. Ocean cultures: Northwest coast ecosystems and indigenous management systems. Chapter 9 in *Conservation for the Anthropocene Ocean*, ed. Phillip S. Levin and Melissa R. Poe. London: Academic Press, 169–206.

Newcombe, C. P., and Macdonald, D. D. 1991. Effects of suspended sediments on aquatic ecosystems. *North American Journal of Fisheries Management* 11: 1, 72–82.

Palmer, A. D. 2005. *Maps of Experience: The Anchoring of Land to Story in Secwepemc Discourse.* Toronto, ON: University of Toronto Press.

Reimchen, T., and Fox, C. H. 2013. Fine-scale spatiotemporal influences of salmon on growth and nitrogen signatures of Sitka spruce tree rings. *BMC Ecology* 13: 1–13.

Ryan, T. 2014. Territorial jurisdiction: The cultural and economic significance of eulachon *Thaleichthys pacificus* in the north-central coast region of British Columbia. PhD dissertation, University of British Columbia. DOI: 10.14288/1.0167417.

Scheffer, M., and Carpenter, S. R. 2003. Catastrophic regime shifts in ecosystems: Linking theory to observation. *Trends in Ecology and Evolution* 18: 648–56.

Simard, S. W. 2016. How trees talk to each other. TED Summit, Banff, AB. https://www .ted.com/talks/suzanne_simard_how_trees_talk_to_each_other?language=en.

Simard, S. W., et al. 2016. From tree to shining tree. *Radiolab* with Robert Krulwich and others. https://www.wnycstudios.org/story/from-tree-to-shining-tree.

Turner, N. J. 2008. Kinship lessons of the birch. *Resurgence* 250: 46–48.

———. 2014. *Ancient Pathways, Ancestral Knowledge: Ethnobotany and Ecological Wisdom of Indigenous Peoples of Northwestern North America.* Montreal, QC: McGill–Queen's Press.

Turner, N. J., Berkes, F., Stephenson, J., and Dick, J. 2013. Blundering intruders: Multiscale impacts on Indigenous food systems. *Human Ecology* 41: 563–74.

Turner, N. J., Ignace, M. B., and Ignace, R. 2000. Traditional ecological knowledge and wisdom of Aboriginal peoples in British Columbia. *Ecological Applications* 10: 1275–87.

White, E. A. F. (Xanius). 2006. Heiltsuk stone fish traps: Products of my ancestors' labour. Master of arts thesis, Simon Fraser University.

EPILOGUE: THE MOTHER TREE PROJECT

Aitken, S. N., and Simard, S. W. 2015. Restoring forests: How we can protect the water we drink and the air we breathe. *Alternatives Journal* 4: 30–35.

Chambers, J. Q., Higuchi, N., Tribuzy, E. S., and Trumbore, S. E. 2001. Carbon sink for a century. *Nature* 410: 429.

Dickinson, R. E., and Cicerone, R. J. 1986. Future global warming from atmospheric trace gases. *Nature* 319: 109–15.

Harris, D. C. 2010. Charles David Keeling and the story of atmospheric CO_2 measurements. *Analytical Chemistry* 82: 7865–70.

Roach, W. J., Simard, S. W., Defrenne, C. E., et al. 2020. Carbon storage, productivity and biodiversity of mature Douglas-fir forests across a climate gradient in British Columbia. (In prep.)

Simard, S. W. 2013. Practicing mindful silviculture in our changing climate. *Silviculture Magazine* Fall 2013: 6–8.

———. 2015. Designing successful forest renewal practices for our changing climate. Natural Sciences and Engineering Council of Canada, Strategic Project Grant. (Proposal for the Mother Tree Project.)

Simard, S. W., Martin, K., Vyse, A., and Larson, B. 2013. Meta-networks of fungi, fauna and flora as agents of complex adaptive systems. Chapter 7 in *Managing World Forests as Complex Adaptive Systems: Building Resilience to the Challenge of Global Change*, ed. K. Puettmann, C. Messier, and K. D. Coates. New York: Routledge, 133–64.

INDEX

Page numbers in *italics* refer to photographs.

Interior

Page 9: Peter Simard; page 10: Sterling Lorence; page 12: Jens Wieting; page 15: Gerald Ferguson; page 22: Winnifred Gardner; page 27: Courtesy of Enderby & District Museum & Archives, EMDS 1430; page 28: Peter Simard; page 29: Courtesy of Enderby & District Museum & Archives, EMDS 1434; page 32: Courtesy of Enderby & District Museum & Archives, EMDS 0541; page 33: Courtesy of Enderby & District Museum & Archives, EMDS 0460; page 34: Courtesy of Enderby & District Museum & Archives, EMDS 0464; page 37: (top) Courtesy of Enderby & District Museum & Archives, EMDS 0461; page 37: (bottom) Courtesy of Enderby & District Museum & Archives, EMDS 0392; page 47: Jean Roach; page 55: Patrick Hattenberger; page 73: Jean Roach; page 86: Jean Roach; page 139: Patrick Hattenberger; page 223: Bill Heath; page 231: Jens Wieting; page 234: Bill Heath; page 243: Bill Heath; page 265: Bill Heath; page 273: Robyn Simard; page 288: Bill Heath; page 295: Emily Kemps; page 301: Bill Heath

Insert 1

Page 1: Jens Wieting; page 2: Jens Wieting; page 3: Jens Wieting; page 4: (top) Bill Heath; page 4: (bottom) Paul Stamets; page 5: Dr. Teresa (Sm'hayetsk) Ryan; page 6: (top) Camille Defrenne; page 6: (bottom) Peter Kennedy, University of Minnesota; page 7: (top) Camille Vernet; page 7: (bottom) Jens Wieting; page 8: Jens Wieting

Insert 2

Page 1: Bill Heath; page 2: Dr. Teresa (Sm'hayetsk) Ryan; page 3: (top) Camille Vernet; page 3: (bottom) Joanne Childs and Colleen Iversen / Oak Ridge National Laboratory, U.S. Department of Energy; page 4: Jens Wieting; page 5: (top and bottom) Jens Wieting; page 6: (top) Paul Stamets; page 6: (bottom) Kevin Beiler; page 7: Dr. Teresa (Sm'hayetsk) Ryan; page 8: Diana Markosian

All other photographs are courtesy of the author.

ALLEN LANE
an imprint of
PENGUIN BOOKS

Also Published

Chiara Marletto, *The Science of Can and Can't: A Physicist's Journey Through the Land of Counterfactuals*

Suzanne Simard, *Finding the Mother Tree: Uncovering the Wisdom and Intelligence of the Forest*

Giles Fraser, *Chosen: Lost and Found between Christianity and Judaism*

Malcolm Gladwell, *The Bomber Mafia: A Story Set in War*

Kate Darling, *The New Breed: How to Think About Robots*

Serhii Plokhy, *Nuclear Folly: A New History of the Cuban Missile Crisis*

Sean McMeekin, *Stalin's War*

Michio Kaku, *The God Equation: The Quest for a Theory of Everything*

Michael Barber, *Accomplishment: How to Achieve Ambitious and Challenging Things*

Charles Townshend, *The Partition: Ireland Divided, 1885-1925*

Hanif Abdurraqib, *A Little Devil in America: In Priase of Black Performance*

Carlo Rovelli, *Helgoland*

Herman Pontzer, *Burn: The Misunderstood Science of Metabolism*

Jordan B. Peterson, *Beyond Order: 12 More Rules for Life*

Bill Gates, *How to Avoid a Climate Disaster: The Solutions We Have and the Breakthroughs We Need*

Kehinde Andrews, *The New Age of Empire: How Racism and Colonialism Still Rule the World*

Veronica O'Keane, *The Rag and Bone Shop: How We Make Memories and Memories Make Us*

Robert Tombs, *This Sovereign Isle: Britain In and Out of Europe*

Mariana Mazzucato, *Mission Economy: A Moonshot Guide to Changing Capitalism*

Frank Wilczek, *Fundamentals: Ten Keys to Reality*

Milo Beckman, *Math Without Numbers*

John Sellars, *The Fourfold Remedy: Epicurus and the Art of Happiness*

T. G. Otte, *Statesman of Europe: A Life of Sir Edward Grey*

Alex Kerr, *Finding the Heart Sutra: Guided by a Magician, an Art Collector and Buddhist Sages from Tibet to Japan*

Edwin Gale, *The Species That Changed Itself: How Prosperity Reshaped Humanity*

Simon Baron-Cohen, *The Pattern Seekers: A New Theory of Human Invention*

Christopher Harding, *The Japanese: A History of Twenty Lives*

Carlo Rovelli, *There Are Places in the World Where Rules Are Less Important Than Kindness*

Ritchie Robertson, *The Enlightenment: The Pursuit of Happiness 1680-1790*

Ivan Krastev, *Is It Tomorrow Yet?: Paradoxes of the Pandemic*

Tim Harper, *Underground Asia: Global Revolutionaries and the Assault on Empire*

John Gray, *Feline Philosophy: Cats and the Meaning of Life*

Priya Satia, *Time's Monster: History, Conscience and Britain's Empire*

Fareed Zakaria, *Ten Lessons for a Post-Pandemic World*

David Sumpter, *The Ten Equations that Rule the World: And How You Can Use Them Too*

Richard J. Evans, *The Hitler Conspiracies: The Third Reich and the Paranoid Imagination*

Fernando Cervantes, *Conquistadores*

John Darwin, *Unlocking the World: Port Cities and Globalization in the Age of Steam, 1830-1930*

Michael Strevens, *The Knowledge Machine: How an Unreasonable Idea Created Modern Science*

Owen Jones, *This Land: The Story of a Movement*

Seb Falk, *The Light Ages: A Medieval Journey of Discovery*

Daniel Yergin, *The New Map: Energy, Climate, and the Clash of Nations*

Michael J. Sandel, *The Tyranny of Merit: What's Become of the Common Good?*

Joseph Henrich, *The Weirdest People in the World: How the West Became Psychologically Peculiar and Particularly Prosperous*

Leonard Mlodinow, *Stephen Hawking: A Memoir of Friendship and Physics*

David Goodhart, *Head Hand Heart: The Struggle for Dignity and Status in the 21st Century*

Claudia Rankine, *Just Us: An American Conversation*

James Rebanks, *English Pastoral: An Inheritance*

Robin Lane Fox, *The Invention of Medicine: From Homer to Hippocrates*

Daniel Lieberman, *Exercised: The Science of Physical Activity, Rest and Health*

Sudhir Hazareesingh, *Black Spartacus: The Epic Life of Touissaint Louverture*

Judith Herrin, *Ravenna: Capital of Empire, Crucible of Europe*

Samantha Cristoforetti, *Diary of an Apprentice Astronaut*

Neil Price, *The Children of Ash and Elm: A History of the Vikings*

George Dyson, *Analogia: The Entangled Destinies of Nature, Human Beings and Machines*

Wolfram Eilenberger, *Time of the Magicians: The Invention of Modern Thought, 1919-1929*

Kate Manne, *Entitled: How Male Privilege Hurts Women*

Christopher de Hamel, *The Book in the Cathedral: The Last Relic of Thomas Becket*

Isabel Wilkerson, *Caste: The International Bestseller*

Bradley Garrett, *Bunker: Building for the End Times*

Katie Mack, *The End of Everything: (Astrophysically Speaking)*

Jonathan C. Slaght, *Owls of the Eastern Ice: The Quest to Find and Save the World's Largest Owl*

Carl T. Bergstrom and Jevin D. West, *Calling Bullshit: The Art of Scepticism in a Data-Driven World*

Paul Collier and John Kay, *Greed Is Dead: Politics After Individualism*

Anne Applebaum, *Twilight of Democracy: The Failure of Politics and the Parting of Friends*

Sarah Stewart Johnson, *The Sirens of Mars: Searching for Life on Another World*